CW00457283

PHYSICS OF FULLY IONIZED GASES

LYMAN SPITZER, JR.

SECOND REVISED EDITION

DOVER PUBLICATIONS, INC.
MINEOLA, NEW YORK

Bibliographical Note

This Dover edition, first published in 2006, is an unabridged republication of the 1962 second revised edition of the work originally published in 1956 by John Wiley and Sons, Inc., New York.

Library of Congress Cataloging-in-Publication Data

Spitzer, Lyman, 1914–
Physics of fully ionized gases / Lyman Spitzer.—2nd rev. ed.
p. cm.
Rev. ed. of: New York : Interscience Publishers, 1956.
Includes bibliographical references and index.
ISBN-13: 978-0-486-44982-1
ISBN-10: 0-486-44982-3 (pbk.)
1. Plasma (Ionized gases) I. Title.

QC718.S6 2006
530.4'4—dc22

2006040275

Manufactured in the United States by LSC Communications
44982304 2019
www.doverpublications.com

Preface to the Second Edition

In the six years since the first edition of this tract was written, research in plasma physics has expanded greatly. Especially since 1958, when research on problems of controlled fusion was declassified, many papers have been published on physical processes in fully ionized gases. Our understanding of many subjects has substantially increased. Two topics where the scale of effort and the increased understanding have been particularly striking are the nature and stability of hydromagnetic equilibria and the propagation of infinitesimal waves through a plasma.

A primary problem in revising this tract has been the selection of topics for inclusion. In the main, the original concept of the tract has been followed and chief emphasis has been placed on the macroscopic equations and their consequences. As in the first edition, however, a preliminary chapter deals with free particle motions, while a final chapter treats encounters between charged particles, giving for reference the various coefficients that must be employed in the macroscopic equations. The chapter on waves has been almost entirely rewritten and considerably expanded, and a new chapter added on hydromagnetic equilibria and their stability. The extensive results on plasma dynamics obtained during the last few years by the use of the Boltzmann equation have been largely ignored. However, the physical mechanisms underlying such central phenomena as Landau damping, cyclotron damping, and two-stream instabilities are treated from a very simple point of view at the end of Chapter 3 on waves.

The chief area of research that has been excluded in the revised tract is the analysis of nonlinear phenomena, such as shocks and turbulence. While several idealized nonlinear problems have been solved recently, our understanding of this subject in general is still very limited.

The extensive burgeoning of plasma research within the last few years is particularly evident in the large number of published papers that should be included in a complete bibliography. While the number of references cited is about twice that of the first edition, no attempt at completeness has been made here; the references listed are primarily those known to the author, and do not necessarily provide either the earliest or the most complete treatment of any subject. As a natural result, the papers listed show a substantial bias in general for U.S. work, and in particular for contributions from the Plasma Physics Laboratory.

It is a pleasure to record my indebtedness to the many scientists who have suggested improvements and corrections to this material, in particular, to T. Northrop for his thoughtful comments on Chapters 1 and 2, to S. J. Buchsbaum, A. F. Kuckes, and M. A. Rothman for detailed suggestions on Chapter 3, to A. N. Kaufman and H. Dreicer for their remarks on Chapter 5, to W. E. Meador for his comments on the Appendix, and to A. Baños, A. Bishop, J. Dawson, M. B. Gottlieb, and W. A. Newcomb for a number of helpful suggestions. My special thanks are due to A. Simon and T. Stix for their critical reading of the entire manuscript.

LYMAN SPITZER, JR.

Plasma Physics Laboratory
Princeton University
Princeton, New Jersey
October, 1961

Preface to the First Edition

Both in gaseous electronics and in theoretical astrophysics there is a growing interest in gases which are almost completely ionized. Although, of course, ionization is never entirely complete, under some conditions the fraction of neutral atoms present may be less than a few per cent, and such atoms may therefore be neglected in discussing most of the physical properties of the gas. Moreover, in the case of hydrogen, which is overwhelmingly the most abundant element in the stars and in space, atoms which are ionized are also stripped. Helium, the next most abundant nucleus, is mostly stripped of its two orbital electrons inside the sun and in the solar corona. Even in a laboratory gas certain observed phenomena, such as plasma oscillations, are independent both of the presence of neutral particles and of the presence of bound electrons in the ionized atoms. Thus for many purposes it is useful to analyze theoretically the behavior of a gas composed entirely of electrons and bare nuclei.

Such a gas has the advantage of considerable simplicity in certain respects. Most quantum-mechanical effects can usually be ignored, except for a relatively weak interaction with the radiation field. Most of the phenomena important in normal gaseous electronics disappear; electron attachment, dissociative recombination, excitation and deexcitation of atoms and molecules, electrical breakdown, etc., do not occur in a fully ionized gas. Since a solid surface would reduce the ionization, any such surface must lie far from the regions being considered, and hence the complicated processes occurring at a solid surface may be ignored. Likewise, the encounters between charged particles become in principle much simpler, as inverse-square forces are more precisely calculable than the complicated interactions of systems containing bound electrons.

The problems encountered in analyzing a fully ionized gas are of several types. Although the basic physical processes are simpler than in an ordinary gas, the motions are more complex, since these are coupled to the electromagnetic field. In the presence of a strong magnetic field this coupling between dynamics and the electromagnetic field gives rise to novel phenomena, first explored by Alfvén, which are included under the general heading of magneto-hydrodynamics, or, more simply, hydromagnetics. The latter term will be employed here. Even in the absence of a magnetic field the electrical properties of a completely ionized gas permit complicated motions, which involve electrostatic restoring forces, and which have no parallel in ordinary gases. Finally, the theory of collisions between particles, in so far as these determine the transport coefficients—electrical and thermal conductivity, viscosity, etc.—and the time of relaxation—the time required to establish an equilibrium velocity distribution—may be approached with a new viewpoint, because of the long-range character of the inverse-square forces involved.

Considerable progress has been made in these fields during the last few years, especially as a result of the work by Alfvén, Cowling, and Schlüter. No general but simple introduction to the subject now exists, and any one wishing to familiarize himself with this area must consult mostly original papers in a variety of journals. The purpose of the present tract is to provide such an introduction, designed for students at the graduate level.

The subject matter is restricted to those topics that may serve to give a theoretical understanding of the subject. Although some observational data are available on certain phases of the subject, as, for example, electromagnetic and electrostatic waves in plasmas, this material has been entirely excluded. To have added a detailed comparison with observations would have meant a considerable increase in the length and scope of this tract.

The book is designed for those who have had an introductory course in theoretical physics, but are otherwise unacquainted with the detailed kinetic theory of gases. For example, a knowledge of Maxwell's equations is assumed, and likewise a familiarity with the elements of vector analysis, such as is provided in the introductory chapter of Page's *Introduction to Theoretical Physics.* The bibliography is by no means complete, but it includes some of the basic papers in each area. It is hoped those who may work in the general field of fully ionized gases will find the references a useful introduction to a new and rapidly growing area of physics.

The author is greatly indebted to M. Savedoff, M. Schwarzschild, A. Simon, T. Stix, and L. Tonks for their careful reading of the manuscript and for a number of important suggestions.

LYMAN SPITZER, JR.

Princeton, New Jersey
July, 1955

Contents

CHAPTER 1

Motion of a Particle

The motion of a charged particle under given external fields has been understood for some time. The analysis in the present introductory chapter, which follows closely the presentation by Alfvén (1), may serve as a review of this field, which is basic in the understanding of dynamical processes in an ionized gas. We shall use electromagnetic units throughout the present tract; to avoid confusion with the standard definition of the electron charge, e, in electrostatic units, the electron charge in electromagnetic units (e.m.u.) will be denoted by $-e/c$ throughout.

1.1 Equations of Motion

When a particle of charge q moves through a region where an electric field $\mathbf{E}$ and a magnetic field $\mathbf{B}$ are present, the particle is subject to two forces. The electrical force, which is parallel to the field $\mathbf{E}$, equals $q\mathbf{E}$ dynes, where both q and $\mathbf{E}$ are measured in electromagnetic units. The magnetic force is at right angles both to $\mathbf{w}$, the velocity of the particle, and $\mathbf{B}$, the magnetic field strength. If $\mathbf{w}$ is measured in c.g.s. units, and $\mathbf{B}$, in gauss, the magnetic force is $q\mathbf{w} \times \mathbf{B}$ dynes. The basic equation of motion is then

$$m \frac{d\mathbf{w}}{dt} = q(\mathbf{E} + \mathbf{w} \times \mathbf{B}) \tag{1-1}$$

where m is the particle mass, in grams.

This familiar equation possesses simple solutions in several special cases. When $\mathbf{B}$ vanishes, and $\mathbf{E}$ is constant in space and time, the particle moves with constant acceleration $q\mathbf{E}/m$. When $\mathbf{E}$ vanishes, the acceleration is $q\mathbf{w} \times \mathbf{B}/m$, and is always perpendicular to the velocity, producing a curvature of the particle path but no change in the scalar velocity w. Thus the kinetic energy of a particle is unaffected by a magnetic field alone. If $\mathbf{E}$ vanishes and $\mathbf{B}$ is constant in space and time, the acceleration is constant in magnitude, and if $\mathbf{w}$ is initially perpendicular to $\mathbf{B}$ the particle will move in a circle of radius a. If the acceleration qwB/m is set numerically equal to the centrifugal acceleration w^2/a, we find at once that the angular frequency w/a is given by

$$\omega_c = \frac{qB}{m} = \frac{ZeB}{mc} \tag{1-2}$$

the particle charge has been set equal to Z times e/c, where e is 4.803×10^{-10}. The quantity ω_c will be called the cyclotron frequency, since it is equal to the angular frequency with which particles gyrate in a cyclotron. For the corresponding frequency, ν_c, in cycles per second, we find

$$\nu_c = \frac{\omega_c}{2\pi} = 1.54 \times 10^3 \frac{ZB}{A} \text{ sec}^{-1} \tag{1-3}$$

where A is the ratio of the particle mass to the mass of unit atomic weight, 1.660×10^{-24} gm. If Z is negative its absolute value must be taken.

The radius of gyration, a, is equal to w/ω_c. If $\mathbf{w}$ is not initially perpendicular to $\mathbf{B}$, the perpendicular component, which we denote by $\mathbf{w}_\perp$, must be used, and a becomes

$$a = \frac{w_\perp}{\omega_c} = \frac{mw_\perp c}{ZeB} \tag{1-4}$$

The component of $\mathbf{w}$ parallel to $\mathbf{B}$, which we denote by $w_\parallel$, will not be affected by the magnetic field, and will have no effect on the motions perpendicular to the field. When these two motions are combined, the final particle path will be a helix of constant pitch around a line of force.

In the following sections the motions of a free particle in other relatively simple cases will be treated. While an understanding of these motions is helpful, it should be emphasized that in the normal ionized gas, where currents and charges in the gas may be of importance, the single-particle picture is frequently not very convenient. Such currents and charges are found much more easily from the macroscopic equations of the gas, developed in the next chapter, than from the microscopic motions of single particles.

1.2 Particle Drifts

We now consider the motion of a charged particle which is moving in a magnetic field $\mathbf{B}$, but is subject to various perturbations, such as the presence of an electric field, or a small spatial inhomogeneity in $\mathbf{B}$, or a slow change of $\mathbf{B}$ in time. In such cases the motion can be described approximately as gyration around a point which is moving. This instantaneous center of gyration is called the "guiding center" of the particle. The motion of the guiding center transverse to $\mathbf{B}$ is called a "drift" of the particle.

a. Electric Field. Let $\mathbf{E}$ and $\mathbf{B}$ be constant in space and time, and let $\mathbf{E}$ be perpendicular to $\mathbf{B}$. We define a new velocity $\mathbf{w}'$ by the condition that

$$\mathbf{w} = \mathbf{w}' + \frac{\mathbf{E} \times \mathbf{B}}{B^2} \tag{1-5}$$

Since $\mathbf{E}$ and $\mathbf{B}$ are assumed constant in both space and time, substitution of equation (1-5) into (1-1) gives

$$m\frac{d\mathbf{w}'}{dt} = q\left\{\mathbf{E} + \mathbf{w}' \times \mathbf{B} + \frac{1}{B^2}(\mathbf{E} \times \mathbf{B}) \times \mathbf{B}\right\} \tag{1-6}$$

If we expand the triple vector product, taking into account that $\mathbf{B}\cdot\mathbf{E}$ is assumed zero, we have

$$(\mathbf{E} \times \mathbf{B}) \times \mathbf{B} = -B^2\mathbf{E} \tag{1-7}$$

Combination of equations (1-6) and (1-7) gives

$$m\frac{d\mathbf{w}'}{dt} = q\mathbf{w}' \times \mathbf{B} \tag{1-8}$$

The motion defined by equation (1-8) is independent of the electric field, and consists of a gyration around the lines of force at the cyclotron frequency. The total velocity $\mathbf{w}$ is the sum of $\mathbf{w}'$ and a drift velocity $\mathbf{w}_D$, perpendicular to both $\mathbf{E}$ and $\mathbf{B}$. In the general case where $\mathbf{E}$ has components $\mathbf{E}_\perp$ and $E_\parallel$, perpendicular and parallel, respectively, to $\mathbf{B}$, this transverse drift velocity is given numerically by

$$w_D = \frac{E_\perp}{B} = \frac{10^8 E_\perp(\text{volt/cm})}{B} \tag{1-9}$$

The component $E_\parallel$ will produce a uniform acceleration along the magnetic lines of force. If the value of w_D computed from equation (1-9) exceeds c, the velocity of light, this equation is, of course, invalid. In this case the kinetic energy of the particle transverse to the magnetic field increases continuously. For a magnetic field of 10^3 gauss, equation (1-9) may be used as long as $E_\perp$ is less than 3×10^{13} e.m.u. or 3×10^5 volts/cm.

The drift velocity given by equation (1-9) may be interpreted in either of two ways. Suppose that a positively charged particle is gyrating perpendicular to the magnetic field, as shown in Figure 1.1. The magnetic field is taken to be directed upwards out of the plane of the paper. If now an electric field is applied, the particle will accelerate on the left-hand side of its circle, and decelerate on the right-hand side; as a result the velocity on the side toward the top of the page will exceed the velocity on the opposite side. According to equation (1-4) the radius of gyration increases with velocity, and hence the radius of curvature of the particle's path is greater on the side near the top of the page than on the other side. A drift to the right results. For particles of opposite sign, the gyration will be in the opposite direction, but the acceleration produced by the electric field will also be reversed, and the drift will

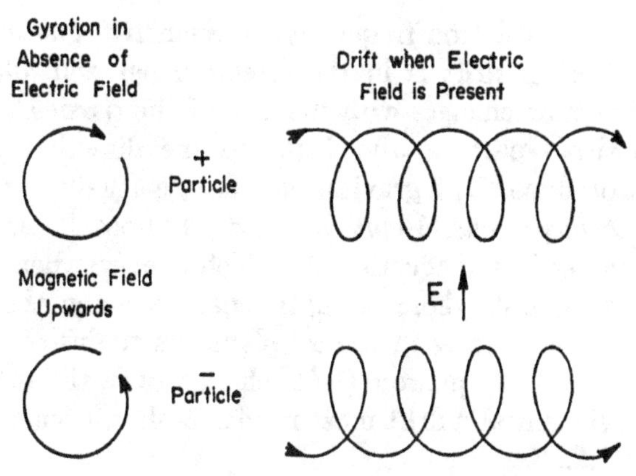

Figure 1.1. Drift produced by an electric field.

be in the same direction as before. Analysis, in these terms, of the magnitude of $\mathbf{w}_D$ shows that it will be independent of both of the particle's mass and its velocity, as well as of the sign of its charge.

A more basic interpretation is in terms of the transformation of **E** and **B** in moving systems. In a system moving with the velocity $\mathbf{E} \times \mathbf{B}/B^2$ there is no electric field, provided that $E_\parallel$ is zero. Thus for an observer moving at the velocity $\mathbf{w}_D$ the electric field transverse to **B** has been transformed away, and as viewed by such an observer the particles must obviously circle around the lines of force. This argument makes it clear that equation (1-9) is valid for particles of relativistic energy, although equation (1-1) is non-relativistic.

b. Gravitational Field. If a particle is subject to a gravitational force which has a component $m\mathbf{g}_\perp$, perpendicular to **B**, a drift will result exactly as in the presence of an electric field. The force per unit charge, which is **E** in the previous case, becomes $m\mathbf{g}_\perp/q$ in the present case, where q is again the charge on the particle. The drift velocity w_D then becomes, on substituting for $E_\perp$ in equation (1-9),

$$w_D = \frac{mg_\perp}{qB} = \frac{g_\perp}{\omega_c} \tag{1-10}$$

where ω_c, the cyclotron frequency, is given by equation (1-2). The direction of drift is in the direction perpendicular to **B** and $\mathbf{g}_\perp$, but now changes with the sign of the particle's charge. For a positive particle, the drift has the direction $\mathbf{g} \times \mathbf{B}$. The drift produced by a gravitational field is usually very small.

c. Inhomogeneous Magnetic Field. Suppose now that a particle moves in a magnetic field which is everywhere parallel to the z axis, but whose strength changes along the x axis. As the particle gyrates in the xy plane, its radius of gyration will, according to equation (1-4), change over the orbit. As in the previous case a drift must result, as shown schematically in Figure 1.2.

In contrast to the cases immediately above, the drift velocity can now be found in general only by means of an approximate theory, in which small terms are ignored. The "first-order" theory, in which only terms of the first order in $w_D/w_\perp$ are retained, has been developed by Alfvén (1). His result, in the present notation, is

$$\frac{w_D}{w_\perp} = \frac{a\nabla_\perp B}{2B} \tag{1-11}$$

where a is again the radius of gyration, given in equation (1-4),

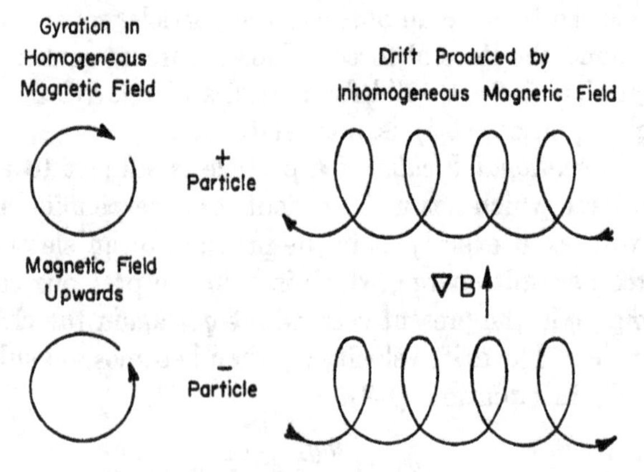

Figure 1.2. Drift produced by an inhomogeneous magnetic field.

and $\nabla_\perp B$ is the gradient of the scalar, B, in the plane perpendicular to $\mathbf{B}$. Apart from the factor 2, equation (1-11) can readily be deduced from dimensional considerations.

Similarly a drift arises if a particle is moving with a velocity $w_\parallel$ along a line of force which is curved with a radius of curvature R. We introduce a new coordinate system rotating with an angular velocity $w_\parallel/R$ about the center of curvature of the field. In this system the particle has no motion along the line of force, but the centrifugal force $mw_\parallel^2/R$ produces the same drift as a gravitational force mg of the same magnitude. Equation (1-10) may be used, and we find

$$w_D = \frac{w_\parallel^2}{R\omega_c} \tag{1-12}$$

If no currents are present in the plasma, $\nabla \times \mathbf{B}$ vanishes, and $(\nabla_\perp B)/B$ equals $1/R$. The sum of equations (1-11) and (1-12) then yields

$$w_D = \frac{1}{\omega_c R}\left(\tfrac{1}{2}w_\perp^2 + w_\parallel^2\right) \tag{1-13}$$

The two drifts are in the same direction; for a positive particle w_D has the direction $\mathbf{B} \times \nabla B$, with the opposite direction for a negative particle.

When the electric and magnetic fields are not constant in time, additional drifts arise in the first-order theory. More specifically, if $\mathbf{E} \times \mathbf{B}/B^2$ changes with time along the path of the guiding center, drifts arise which are first order in $1/\omega_c$, but of higher order in $1/B$. A systematic discussion of all these drifts to first order in $1/\omega_c$ has been given by Northrop (7).

d. Confinement in Axial Fields. In the general case the drifts discussed above naturally affect the confinement of charged particles in a magnetic field, a topic of interest both in astronomical and in terrestrial plasmas. However, if the electric and magnetic fields are axially symmetric, and if the kinetic energy is not too great, an exact proof may be given that these drifts do not affect particle confinement; specifically, a charged particle will always oscillate about some surface of constant flux, generated by rotating some line of magnetic force about the axis of symmetry.

To establish this result, we use the θ component of equation (1-1) for the acceleration around the axis of symmetry. Since the electric potential U, like the magnetic field $\mathbf{B}$, is assumed independent of θ, there can be no electrostatic contribution to E_θ. If we also assume that $\mathbf{B}$ is constant with time, there can be no induced $\mathbf{E}$, and E_θ vanishes. To simplify the term $w_z B_r - w_r B_z$ we introduce $\Phi(r, z)$, the magnetic flux through a circle of radius r in a plane of given z. The defining relation for Φ in differential form is

$$\frac{\partial \Phi}{\partial r} = 2\pi r B_z \tag{1-14}$$

From the condition that $\nabla \cdot \mathbf{B}$ is zero we find, letting $\partial B_\theta / \partial \theta$ equal zero,

$$\frac{\partial \Phi}{\partial z} = -2\pi r B_r \tag{1-15}$$

The θ component of equation (1-1), relating the change of angular momentum to the torque, can now be integrated directly; we obtain

$$mrw_\theta + \frac{q}{2\pi} \Phi(r, z) = C \tag{1-16}$$

where C is some constant. The magnetic field component B_θ may be an arbitrary function of r and z without affecting this result.

It is readily shown from Maxwell's equations that if $\mathbf{B}$ is a function of time, $2\pi r E_\theta$ equals $-\partial \Phi / \partial t$, and equation (1-16) is still valid, except that Φ is now a function of t, as well as of r and z. From electrodynamic theory it is evident that $\Phi / 2\pi r$ is the θ component of the usual vector potential and that equation (1-16) is the familiar constancy of the generalized angular momentum in the absence of external torques. This result is exact also for relativistic velocities, if m is taken as the relativistic transverse mass rather than the rest mass.

Equation (1-16) may be used to demonstrate particle confinement generally. We restrict ourselves here, for simplicity, to trajectories which do not enclose the axis of symmetry and for which, as a result, w_θ vanishes twice during each gyration. When w_θ is zero, equation (1-16) gives a simple relationship between r and z; we denote by $r_1(z, C)$ the function r_1 determined in this way. Since Φ is defined as the magnetic flux crossing a circle of radius r, it is clear that Φ is constant along any line of force. Hence the points at which w_θ vanishes lie on a surface generated by rotating some line of force about the axis of symmetry; we call this surface a "surface of constant flux." Different such surfaces are distinguished by different values of the constant, C. As the particle gyrates it continues to cross back and forth across the constant flux surface $r_1(z, C)$.

Physically, one would expect the excursions of a charged particle from this surface to be relatively small if a, the radius of gyration, is small compared to the axial distance, r. Mathematically, w_θ^2 has at each point a maximum value determined by the known particle energy and the electric potential, $U(r, z)$. This upper limit on w_θ yields, from equation (1-16), an upper limit on $r - r_1(z, C)$, provided that Φ is a monotonic function of r, with a sufficient range of variation. The particle never departs from the surface of constant flux by more than this upper limit, which is about equal to the radius of gyration if Φ changes nearly linearly over this range of r. In this sense perfect confinement of a single particle against radial loss is assured in any axisymmetric system, if there are no collisions.

1.3 Magnetic Moment

For slow variations of **B** in space and time the diamagnetic moment, μ, of a charged particle is nearly constant, and provides an approximate integral of the motion. The magnetic moment of a current I encircling an area S equals IS. In the present instance S is simply πa^2, where a is the radius of gyration. The current equals the charge q multiplied by $\omega_c/2\pi$, the number of gyrations per second. Hence we have

$$\mu = \pi a^2 \cdot \frac{q\omega_c}{2\pi} = \frac{\frac{1}{2}mw_\perp{}^2}{B} \tag{1-17}$$

where we have made use of equations (1-2) and (1-4). The magnetic flux through the particle orbit is directly proportional to μ, since ω_c is directly proportional to B.

Let us consider how μ changes when $\mathbf{B}$ changes with time, but is uniform throughout space. The change of $\mathbf{B}$ will induce an electromotive force around the orbit of the particle. From Faraday's law (see equation (2-18) in the following chapter) we have, using Stokes' theorem,

$$\text{E.M.F.} = \oint \mathbf{E} \cdot \mathbf{ds} = -\int \frac{d\mathbf{B}}{dt} \cdot \mathbf{dS} \tag{1-18}$$

where $\mathbf{ds}$ is a line element around the path, and $\mathbf{dS}$ is an element of the surface enclosed by the path. The change of kinetic energy per unit time is the product of the E.M.F. and the effective current, I; as we have seen, I is $q\omega_c/2\pi$. It is readily shown that the current and the E.M.F. are in the same direction if B is increasing.

Hence

$$\frac{d}{dt}\left(\tfrac{1}{2}mw_\perp{}^2\right) = \frac{q\omega_c}{2\pi} \cdot \pi a^2 \cdot \frac{dB}{dt} = \mu \frac{dB}{dt} \tag{1-19}$$

The rate of change of μ may be found from equation (1-17); on multiplication through by B, and differentiation with respect to time, we find

$$\frac{d}{dt}(\mu B) = \frac{d}{dt}\left(\tfrac{1}{2}mw_\perp{}^2\right) \tag{1-20}$$

Combination of equations (1-19) and (1-20) shows that $d\mu/dt$ vanishes. With μ defined in equation (1-17) this result is valid only for nonrelativistic energies. In the relativistic case, considered by Hellwig (5) and by Vandervoort (9), μ is constant if equation (1-17) is modified in two ways; m must be replaced by m^2/m_0, where m_0 is the rest mass, and both $w_\perp^2$ and B must be measured in a reference frame such that $E_\perp$ vanishes.

A quantity which, like μ, is constant for slow changes of the electric and magnetic fields, is called an "adiabatic invariant." In general, in any periodic motion with one degree of freedom the integral of pdq around the orbit is an adiabatic invariant; p is the generalized momentum associated with the coordinate q. Another important type of adiabatic invariant is discussed in the next section.

The constancy of μ would be exact if the electron charge were distributed uniformly around its circle of gyration. Whether μ tends to be constant depends on the rate at which $\mathbf{B}$ changes. It is obvious physically that if all the change in $\mathbf{B}$ occurs while the electron is moving over a small arc of its circle of gyration, the line integral of $\mathbf{E}$ around the circle, which we have used in equation (1-18), is irrelevant and $d\mu/dt$ does not vanish. However, if we assume that dB/dt is proportional to ωB, and solve the equations of motion to first order in ω/ω_c, then to this order $d\mu/dt$ does in fact vanish.

The extent to which an adiabatic invariant such as the magnetic moment, μ, is constant to higher orders of ω/ω_c has been extensively analyzed. Kruskal (6) has shown that effectively $\Delta\mu$, the change of μ resulting from a variation of $\mathbf{B}$, is zero to all orders of ω/ω_c. To obtain this result during the time that $\mathbf{B}$ is changing requires that the definition of μ in terms of the instantaneous $\mathbf{w}$ and $\mathbf{B}$ be modified to include higher order terms in ω/ω_c. If μ is determined when $\partial\mathbf{B}/\partial t$ and all higher derivatives are zero, these terms disappear from the definition. These results refer to the asymptotic expansion of $\Delta\mu$ in a series of ascending powers of ω/ω_c, and do not imply that $\Delta\mu$ is rigorously zero. A variation of log $(\Delta\mu)$ as $-\omega_c/\omega$ would be consistent with an asymptotic expansion vanishing to all orders of ω/ω_c.

Next we consider the change of μ when **B** varies along the particle path, but is constant with time at each point. Let us suppose that the gyrating particle is moving into a region of greater field. In such a case the lines of force will be convergent, and the magnetic field will have a component B_r directed toward the line of force along which the guiding center is moving (see Figure 1.3). This component produces a retarding force in the direction of the particle's motion.

The magnitude of this force is readily computed. Following Alfvén, we shall simplify the analysis by considering an axisymmetric field, with the guiding center moving along the z axis, taken to be the axis of symmetry. In more complicated situations other drifts may appear, but the essential results obtained here are unchanged. In cylindrical coordinates r, θ, z, the magnetic field is independent of θ; the condition

$$\nabla \cdot \mathbf{B} = 0 \tag{1-21}$$

gives

$$\frac{1}{r}\frac{\partial}{\partial r}(rB_r) + \frac{\partial B_z}{\partial z} = 0 \tag{1-22}$$

If we assume that $\partial B_z/\partial z$ is constant over the cross section of the particle's orbit, and essentially equal to $\partial B/\partial z$, we may integrate equation (1-22) over r to find

$$B_r = -\tfrac{1}{2}r\frac{\partial B}{\partial z} \tag{1-23}$$

Setting r equal to the radius of gyration a, and taking the z component of equation (1-1) we obtain, with use of equation (1-17)

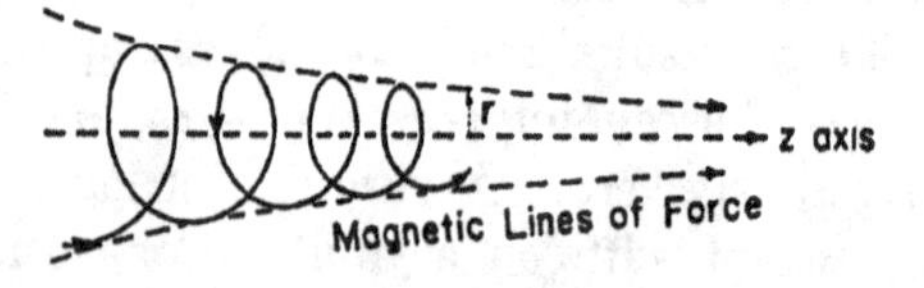

Figure 1.3. Motion of a particle in converging lines of magnetic force.

$$m \frac{dw_{\parallel}}{dt} = -\mu \nabla_{\parallel} B \tag{1-24}$$

where we use the symbol $\nabla_{\parallel}$ to denote the component of the gradient in the direction of **B**. Equation (1-24) is exactly what one would anticipate for a diamagnetic particle.

From equation (1-24) and the conservation of the kinetic energy $\frac{1}{2}m(w_{\perp}^2 + w_{\parallel}^2)$ we can deduce the variation of μ with position. On multiplication of equation (1-24) by $w_{\parallel}$, we obtain

$$\frac{d}{dt}\left(\tfrac{1}{2}mw_{\parallel}^2\right) = -\mu \frac{dB}{dt} \tag{1-25}$$

where d/dt represents the time derivative taken along the path of the particle. By the conservation of energy, and equation (1-17), we have

$$\frac{d}{dt}\left(\tfrac{1}{2}mw_{\parallel}^2\right) = -\frac{d}{dt}\left(\tfrac{1}{2}mw_{\perp}^2\right) = -\frac{d}{dt}(\mu B) \tag{1-26}$$

From equations (1-25) and (1-26) it follows again that μ is a constant of the motion, a conclusion valid here for particles of relativistic energies. Again, this result is approximate and does not hold if B changes markedly over a distance equal to the radius of gyration. If the spatial derivative of **B** is proportional to κB, Kruskal's analysis (6) demonstrates that μ is constant to all orders of $a\kappa$. Again, this result is valid only if the definition of μ is modified to include higher order terms in $a\kappa$.

Evidently the change of μ tends to be very small whenever **E** and **B** change sufficiently slowly in both space and time. When μ can be assumed constant, the motion of a particle in a magnetic field is much simplified, since only the motion of the guiding center need be considered. If two particles have the same guiding center, the same kinetic energy and the same magnetic moment, their guiding centers will then have identical trajectories independent of what phase the particles may have in their gyration around the guiding centers. This convenient result holds to the same accuracy as the constancy of the magnetic moment.

Another important result which follows from the invariance of μ is the reflection of gyrating particles from regions of increasing magnetic field. If θ is the angle which the velocity vector makes with the z axis, then the ratio of $w_\perp$ to the total velocity will be $\sin\theta$. Let θ_0 be the initial value of θ, where B equals B_0. Then as B increases, the constancy of the magnetic moment, $mw_\perp^2/2B$, implies that $w_\perp^2$ increases proportionally to B, and evidently

$$\sin^2\theta = \frac{B}{B_0}\sin^2\theta_0 \tag{1-27}$$

When B/B_0 rises to $1/\sin^2\theta_0$, then all the energy of the particle has been transformed into transverse kinetic energy, $w_\parallel$ falls to zero, and the particle is then reflected back into the region of lesser field. Conversely, if B_m is the maximum value reached by the magnetic field, all particles will be reflected for which $\sin^2\theta_0$ exceeds B_0/B_m. Such a reflecting region may be called a "magnetic mirror." Confinement of charged particles between magnetic mirrors is observed in laboratory devices and in the earth's magnetic field, where the trapped ions above the atmosphere constitute the Van Allen radiation belts.

If we assume an isotropic velocity distribution for the particles within a mirror, we may readily compute the coefficient of reflection, R, defined as the fraction of particles reaching the mirror per unit time that are reflected. Consider particles with some given total initial velocity, w. The number of particles reaching the mirror per second, in the interval $d\Omega$ of solid angle, will be proportional to $\cos\theta_0 d\Omega$. Hence we have

$$R = \int_{\theta_0=\theta_1}^{\pi/2} \cos\theta_0 \, d\Omega \Bigg/ \int_{\theta_0=0}^{\pi/2} \cos\theta_0 \, d\Omega \tag{1-28}$$

where $\sin^2\theta_1$ equals B_0/B_m. Since $d\Omega$ equals $2\pi\sin\theta_0 d\theta_0$, we obtain

$$R = 1 - \frac{B_0}{B_m} \tag{1-29}$$

Since R does not depend on the assumed initial velocity, the same coefficient of reflection applies for an arbitrary velocity distribution, provided only that the distribution is isotropic.

For a trapped particle oscillating between two magnetic mirrors, the velocity parallel to the magnetic field gives rise to the "longitudinal" adiabatic invariant. If we denote by ds the distance interval along the magnetic field, this invariant is the integral of $w_{\parallel} ds$ over one period of oscillation back and forth between the mirrors. As we shall see in the next section, this invariant remains constant as the distance between mirrors changes slowly. In addition, it remains constant if the particle orbit slowly drifts to a different line of force, where the distance, L, between reflection points may be different. If this longitudinal adiabatic is expressed asymptotically in a series of ascending powers of dL/dt, the analyses by Gardner (4) and Kruskal (6) establish constancy to all orders, exactly as in the case of the magnetic moment.

1.4 Acceleration of Particles

The acceleration of charged particles to very high energies is a problem of interest in the study both of cosmic rays and of hot plasmas. Basically, such acceleration requires an electric field. We review briefly here three simple ways in which acceleration can be produced.

In principle the simplest method is acceleration by an electric field in the absence of a magnetic field, or parallel to the magnetic field. If $\mathbf{w} \times \mathbf{B}$ is negligible, the solution of equation (1-1) is trivial. However, in an ionized plasma at high temperature the electrical conductivity is very high, and any electric field parallel to $\mathbf{B}$ is likely to be very small. Hence Fermi (3) and Alfvén (2) have proposed two other methods for accelerating particles to cosmic ray energies; variants of these methods have been proposed for terrestrial plasmas.

In the mechanism proposed by Fermi, a charged particle is moving in a magnetic field between two interstellar clouds. If the magnetic field in the clouds is assumed to be greater than in the intervening region, the particle is trapped between two magnetic mirrors, of the type described in the previous section. Such trapping will occur, of course, only for particles whose velocity is inclined to the magnetic field at a sufficient angle. The clouds comprising the two mirrors are assumed to be moving toward each other at the relative velocity V. A charged particle now gains energy on each reflection from the mirror.

The acceleration may be computed from the constancy of the longitudinal adiabatic invariant discussed in the previous section. We demonstrate here the constancy of this invariant with the simplifying assumption that the magnitude of **B** is constant between the clouds and does not change as the two clouds approach each other. Also, we assume that **B** is axially symmetric around the line joining the two clouds; as shown in Section 1.2, the particle drifts do not change the axial distance in this case and may be ignored. In this simple case $w_{\parallel}$ is constant over virtually the entire path of integration and the longitudinal invariant equals $2w_{\parallel}L$.

To demonstrate that $w_{\parallel}L$ is constant we take one cloud to be stationary, with the other cloud approaching it at a velocity V; then each reflection from the moving cloud will increase the particle velocity by $2V$. The number of such reflections per second will be $w_{\parallel}/2L$, and we have

$$\frac{dw_{\parallel}}{dt} = \frac{w_{\parallel}}{2L} \cdot 2V = -\frac{w_{\parallel}}{L}\frac{dL}{dt} \tag{1-30}$$

Integrating this equation we see that $w_{\parallel}L$ is constant. The corresponding direct proof that the longitudinal invariant is in fact constant in the more general case has been given by Northrop and Teller (8).

It is convenient to express $w_{\parallel}^2$, or the parallel temperature $T_{\parallel}$, in terms of n and B instead of L. Since the total number of trapped particles within a tube of force of radius r is constant, nLr^2 must be constant; since Br^2 is also constant with time we have

$$L \propto \frac{B}{n} \tag{1-31}$$

Hence the kinetic temperature, $T_{\parallel}$, for a group of such particles, defined in terms of the energy $\frac{1}{2}mw_{\parallel}^2$, varies with time according to the law

$$T_{\parallel} \propto \left(\frac{n}{B}\right)^2 \tag{1-32}$$

In the simple case considered above, B is assumed constant, L varies inversely with n, and $T_{\parallel}$ varies as n^2. This proportionality of $T_{\parallel}$ to n^2 is a direct result of the assumption that we have a one-dimensional system; compression is assumed in one dimension, and transfer of kinetic energy to the other two dimensions is neglected. It is well known that for adiabatic compression

$$T \propto n^{\gamma-1} \tag{1-33}$$

where γ, the ratio of the specific heat at constant pressure to that at constant volume, is given by

$$\gamma = \frac{2+m}{m} \tag{1-34}$$

the quantity m is here the number of degrees of freedom. In a fully ionized gas, no internal degrees of freedom need be considered, and for compression in one dimension $m = 1$, $\gamma = 3$, and we arrive at equation (1-32). If the particle velocities were randomized by collisions during the compression, then γ should be set equal to its usual value of 5/3 in equation (1-33).

This method of particle acceleration is subject to one important limitation. As $w_{\parallel}$ increases, the angle, θ, between $\mathbf{w}$ and $\mathbf{B}$ decreases, and the particle is ultimately no longer trapped. Thus the ratio of the total energy to the transverse energy is increased to a certain limit, this limit depending on the reflection coefficient of the magnetic mirrors. Another manifestation of this same difficulty is that the transverse velocity can never be increased as long as B is unchanged, since the magnetic moment, μ, is constant in the absence of collisions and other perturbations. To obtain continuous acceleration of particles one must therefore assume that collisions or other effects re-establish an isotropic velocity distribution, after $w_{\parallel}$ has been increased, and that the particles then become trapped and accelerated again. Since encounters with electrons and positive ions are relatively ineffective for very energetic particles, Fermi suggests that shock waves or plasma oscillations may tend to re-establish an isotropic velocity distribution in interstellar space.

In another mechanism, proposed by Alfvén, charged particles in space are accelerated directly by an increase in the magnetic field. Let us consider a region in which the magnetic field is spatially uniform but increasing in time; from the constancy of μ, the magnetic moment, it is evident that for the particles in this region

$$T_{\perp} \propto B \tag{1-35}$$

where we have used the definition of μ in equation (1-14). Apart from the small drifts discussed in Section 1.2, the charged particles tend to follow the lines of force; as B increases, the lines of force crowd closer together, and in the special case that the compression is strictly two-dimensional n varies as B and hence as $T_{\perp}$. This proportionality between $T_{\perp}$ and n may also be deduced from equations (1-33) and (1-34), since in this case the compression is two-dimensional as far as velocities are concerned, and γ equals 2.

Acceleration of particles in this way also is limited, since the relative change of $T_{\perp}$ equals the relative change in B. Alfvén suggests that particles may pass repeatedly through regions where B is varying with time, and experience repeated accelerations, with collisions or other perturbations reestablishing an isotropic velocity distribution between the periods of acceleration.

References

1. Alfvén, H., *Cosmical Electrodynamics*, Clarendon Press, Oxford, 1950.
2. Alfvén, H., *Tellus*, 6, 232 (1954).
3. Fermi, E., *Astrophys. J.*, 119, 1 (1954).
4. Gardner, C. S., *Phys. Rev.*, 115, 791 (1959).
5. Hellwig, G., *Z. Naturforsch.*, 10a, 508 (1955).
6. Kruskal, M. D., Rendiconti del Terzo Congresso Internazionale sul Fenomeni d'Ionizzazione nei Gas, p. 562 (Societa Italiani di Fisica, Milan, 1957); *Nuclear Fusion Supplement*, in press.
7. Northrop, T. G., *Ann. Phys.*, 15, 79 (1961).
8. Northrop, T. G., and E. Teller, *Phys. Rev.*, 117, 215 (1960).
9. Vandervoort, P., *Ann. Phys.*, 10, 401 (1960).

CHAPTER 2

Macroscopic Behaviour of a Plasma

The study of individual particles frequently gives insight into the behaviour of an ionized gas. However, such a study is not usually the most convenient method for obtaining quantitative information on specific problems. This is partly because the current density **j** plays an important part in most situations, giving rise to both electric and magnetic fields. In the presence of a magnetic field the relationship between **j** and the particle velocities is not a simple one, as we shall see below. Moreover, for any accurate computations a distribution of particle velocities must be taken into account. As a result, the computation of **j** from the velocities of single particles requires a consideration of a discouragingly large number of particles. For rigorous results in complicated situations such considerations cannot be avoided. For rapid but approximate results many specific problems are best analyzed in terms of the macroscopic equations of motion. These equations, together with other needed relationships, are presented in the following sections.

It is surprising, perhaps, that the macroscopic equations presented below do not depend very sensitively on the ratio of the collision frequency ν to the electron cyclotron frequency ω_{ce}. Evidently this ratio has a very great effect on the types of motions of the individual particles. In addition, this ratio affects both the magnitude of the electric current flowing in response to an applied electric field, see Section 2.4, and the proper value of η to use for a current transverse to a magnetic field, see Section 5.4. However, if the resistivity η is small, the macroscopic motion of a plasma is remarkably independent of ν/ω_{ce}, especially if conditions are uniform along each line of force.

To clarify the significance of the basic equations a number of special topics are discussed at the end of this chapter. The subsequent two chapters make detailed application of these equations to some of the standard problems of plasma physics.

2.1 Electrical Neutrality

Before considering the macroscopic equations, we consider first a basic property of a plasma, its tendency toward electrical neutrality. If over a large volume the number of electrons per cubic centimeter deviates appreciably from the corresponding number of positive ions, the electrostatic forces resulting yield a potential energy per particle that is enormously greater than the mean thermal energy. Unless very special mechanisms are involved to support such large potentials, the charged particles will rapidly move in such a way as to reduce these potential differences, i.e., to restore electrical neutrality.

To discuss this problem quantitatively we shall consider a situation where the electric field is everywhere parallel to the x axis. Let us consider that in a certain region no positive ions are present. The electrical potential U is then determined by Poisson's law, equation (2-16), which becomes

$$\frac{d^2U}{dx^2} = 4\pi n_e ec \tag{2-1}$$

Again U is in e.m.u., while the electron charge in e.m.u. is denoted by $-e/c$; n_e denotes the number of electrons per cubic centimeter. If W denotes the potential energy of an electron, equal to $-eU/c$, then the change of W across a slab of width x is given by

$$\Delta W = -2\pi n_e e^2 x^2 \tag{2-2}$$

provided that the electric field vanishes on one side of the slab. We shall denote by the symbol h the value of x for which the absolute value of ΔW equals $\frac{1}{2}kT$, the mean kinetic energy per particle on one direction; T denotes the kinetic temperature in degrees Kelvin, while k is the familiar Boltzmann constant. Evidently

$$h = \left(\frac{kT}{4\pi n_e e^2}\right)^{1/2} = 6.90 \left(\frac{T}{n_e}\right)^{1/2} \tag{2-3}$$

This quantity h, defined by equation (2-3), is called the "Debye shielding distance," since Debye has shown, on the basis of certain approximations, that the field of a point charge in an electrolyte varies as $(1/r) \exp(-r/h)$; at distances r greater than h the electric field of the charge is shielded by particles of opposite sign. Although the precise applicability of Debye's result to an ionized gas is open to question, the Debye shielding distance is clearly a measure of the distance over which n_e can deviate appreciably from $n_i Z$. In Debye's original analysis $n_e + n_i Z^2$ replaces n_e in the denominator of equation (2-3). For shielding of a stationary charge both ions and electrons are effective, and this substitution is an appropriate one. However, Debye's analysis is a very approximate one, and in a more realistic theory ions will certainly have different shielding effects from electrons, especially where rapidly fluctuating phenomena are involved. To consider such effects with any precision requires very detailed study, and for simplicity we may take equation (2-3) as giving a rough measure of the distance over which n_e can deviate appreciably from $n_i Z$. For example, over a region whose thickness is ten times h, the electron density must equal $n_i Z$ to within one per cent, if the electrical potential energy per electron is not to exceed the mean thermal energy. Consideration of three-dimensional geometries does not change the order of magnitude of this result.

If h is small compared with the other lengths of interest, an ionized gas is called a plasma, in accordance with the definition introduced by Langmuir (6).

In plasmas produced within the laboratory the Debye shielding distance is important in that it gives roughly the thickness of the sheath which develops wherever the plasma is in contact with a solid surface. Without such a sheath a plasma, in the absence of a magnetic field, would lose electrons much more rapidly than positive ions, owing to the greater electron velocity. If the potential of the solid surface is allowed to float, no current must flow from the plasma to the surface. In equilibrium a potential gradient arises near the wall, reflecting most of the electrons back into the plasma, the number striking the wall being equal to the corresponding number of positive ions reaching the wall. Within the sheath electrical neutrality is not preserved, even approximately, and eU/c changes through the sheath by an amount comparable with kT. It follows at once that the thickness of the sheath must be about equal to the Debye shielding distance. Detailed physical analyses of a plasma sheath with various assumptions as to the potential difference between the plasma and the wall have been given by Tonks and Langmuir (14).

2.2 Basic Equations

The macroscopic quantities $\mathbf{j}$ and $\mathbf{v}$ are determined by the macroscopic equations of motion, the so-called transfer (or transport) equations of kinetic theory. In view of the basic importance of these equations, their derivation from the Boltzmann equation is given in the Appendix. For ions of charge Ze/c, mass m_i and particle density n_i the equation of motion (6-16) becomes

$$n_i m_i\left(\frac{\partial \mathbf{v}_i}{\partial t} + \mathbf{v}_i \cdot \nabla \mathbf{v}_i\right) = \frac{n_i Z e}{c}(\mathbf{E} + \mathbf{v}_i \times \mathbf{B}) - \nabla \cdot \mathbf{\Psi}_i - n_i m_i \nabla \phi + \mathbf{P}_{ie} \tag{2-4}$$

where ϕ is the gravitational potential, and $\mathbf{v}_i$, the mean velocity of the particles in an element of volume ΔV, is given by

$$\mathbf{v}_i = \frac{1}{n_i \Delta V} \Sigma \mathbf{w}_i \tag{2-5}$$

The quantity $\boldsymbol{\Psi}_i$ is the stress tensor, or dyadic, defined by

$$\boldsymbol{\Psi}_i = \frac{m_i}{\Delta V} \Sigma(\mathbf{w}_i - \mathbf{v}_i)(\mathbf{w}_i - \mathbf{v}_i) \tag{2-6}$$

the summation extending again over the volume element. The quantity $\mathbf{P}_{ie}$ is the total momentum transferred to the ions per unit volume per unit time by collisions with the electrons. If ions of other types are present, the transfer of momentum in encounters with these other particles must also be included. Here we shall assume, for simplicity, that only electrons and one type of positive ion are present. The equation of motion for electrons is obtained from equation (2-4), with $\mathbf{v}_e$, n_e, m_e, $\boldsymbol{\Psi}_e$ and $\mathbf{P}_{ei}$ replacing $\mathbf{v}_i$, n_i, m_i, $\boldsymbol{\Psi}_i$, and $\mathbf{P}_{ie}$, and with Z set equal to -1.

Equation (2-4) is exact for a non-relativistic perfect gas. This equation is useful, however, only in those cases where the distribution of random velocities is sufficiently well behaved so that the stress tensor $\boldsymbol{\Psi}$ may be approximated in a relatively simple way. In general, $\boldsymbol{\Psi}$ has nine components Ψ_{lm}, where l and m represent directions along each of the three coordinate axes. Since Ψ_{lm} equals Ψ_{ml}, there are only six independent components. In the simplest case the distribution of the random velocities $\mathbf{w}_i - \mathbf{v}_i$ is isotropic, Ψ_{ml} vanishes unless m equals l, and the three "diagonal components" Ψ_{xx}, Ψ_{yy}, Ψ_{zz} are all equal to each other and to the scalar pressure, p. In this situation we have

$$\nabla \cdot \boldsymbol{\Psi} = \nabla p \tag{2-7}$$

There are two circumstances in which equation (2-7) is valid. The first arises when the mean free path for collisions between particles is short compared to the distance over which p, $\mathbf{v}$ and other macroscopic quantities change significantly. This is the familiar situation in fluid dynamics, and applies to plasmas within stars, for example. It is well known that the velocity distribution is nearly isotropic when the mean free path is sufficiently short.

With a slight modification equation (2-7) is also valid, even for long mean free paths, provided two other conditions are both satisfied: (a), the plasma is in a magnetic field sufficiently strong so that the gyration radius, a, of each particle is short compared to the distance over which all the macroscopic quantities change appreciably; (b) the plasma configuration and its change with time are such that all gradients along the lines of magnetic force may be neglected. As Watson (15) and Chew, Goldberger and Low (3) have shown, condition (a) leads to the results that Ψ is essentially diagonal, with the two components perpendicular to **B** equal to each other. We denote these components by $p_\perp$; the component parallel to **B** is denoted by $p_\parallel$. This result is physically reasonable in the limit of infinite mean free path, since each individual particle gyrates around its guiding center with a nearly constant velocity, $w_\perp$. Provided that the distribution of guiding centers is nearly uniform over a distance equal to the gyration radius, the disperson of velocities in each of the two directions perpendicular to **B**, resulting from particles of circular velocity $w_\perp$, should be closely equal to $\frac{1}{2}w_\perp^2$.

In the general case the restriction of Ψ to only two independent components, $p_\perp$ and $p_\parallel$, does not suffice to determine a solution of the macroscopic equations. Even in a stationary state both $p_\perp$ and $p_\parallel$ may vary in the direction along **B**. When conditions change with time, determination of the way in which $p_\perp$ and $p_\parallel$ change is not possible in any simple way, since heat energy may flow along the lines of force. When collisions are infrequent, such a heat flow depends on the detailed nature of the velocity distribution function, and cannot be determined in any simple way from the macroscopic equations. If condition (b) is satisfied, in addition to condition (a), then $p_\perp$ and $p_\parallel$ are each constant along a line of force, there is no heat flow parallel to **B**, and the change of $p_\perp$ with changing B is given by the familiar adiabatic law for two-dimensional compression, with γ equal to 2—see Section 1.4. The longitudinal pressure gradient, $p_\parallel$, is now irrelevant, since its gradient along **B** is assumed zero, and if collisions are unimportant equation (2-7) is applicable provided we replace p in equation (2-7) by $p_\perp$. Condition (b) is essentially equivalent to requiring that the problem is two-dimensional, with all conditions uniform along **B**.

In any case the stress tensor will have nondiagonal components Ψ_{xy}, Ψ_{yz}, and Ψ_{zx}, which give rise to viscous stress and which are ignored in equation (2-7). In the limit of short mean free path the assumption that these stresses are linear functions of the quantities $\partial v_r/\partial x_s$, where r and s represent coordinate directions, leads to the usual Navier-Stokes equation for a viscous gas. The effect of viscosity in the opposite limit of long mean free path is discussed briefly in Sections 2.5 and 5.5. Since relatively few situations have been studied where the viscosity of a fully ionized gas is important, we shall neglect these nondiagonal components in the general plasma equations derived here.

In summary, we shall here adopt equation (2-7) as a basic simplification. The macroscopic equations obtained in this way can be applied with high precision either to the case of short mean free path, when p is evidently nearly isotropic, or to the case of a strong, essentially two-dimensional magnetic field, where we may replace p by $p_\perp$. A somewhat special situation in which equation (2-7) also gives reliable results is discussed in Section 3.2 in connection with the propagation of electrostatic waves. In other situations this equation may give a useful indication of what may happen, but should be employed with caution.

From equation (2-4), and the corresponding equation for electrons, one may obtain equations for the macroscopic quantities $\mathbf{v}$ and $\mathbf{j}$. For a gas containing only electrons and one type of positive ion, these quantities are defined by

$$\mathbf{v} = \frac{1}{\rho}(n_i m_i \mathbf{v}_i + n_e m_e \mathbf{v}_e) \tag{2-8}$$

$$\mathbf{j} = \frac{e}{c}(n_i Z \mathbf{v}_i - n_e \mathbf{v}_e) \tag{2-9}$$

where ρ, the mass density, is given by

$$\rho = n_i m_i + n_e m_e \tag{2-10}$$

In analyzing plasma behaviour one may use the macroscopic equations for the ion and electron velocities, $\mathbf{v}_i$ and $\mathbf{v}_e$, or alternatively one may use the equations for $\mathbf{v}$ and $\mathbf{j}$. The former approach has the advantage that it provides a simple, clear derivation of plasma motions in those idealized cases where either the electrons or the positive ions remain at rest. However, in more general cases the use of $\mathbf{v}_e$ and $\mathbf{v}_i$ is somewhat cumbersome, as the current density, $\mathbf{j}$, which influences $\mathbf{E}$ and $\mathbf{B}$, must be determined separately from equation (2-9). To make possible a unified treatment of the subject as a whole we shall utilize the equations for $\mathbf{v}$ and $\mathbf{j}$ through the present tract.

In their exact form, discussed by Schlüter (10) and Lüst (7), these equations are rather complicated. Since these full equations are not needed for any of the subsequent discussion of plasma problems, we shall simplify the analysis with the following three basic approximations:

1. We neglect all the quadratic terms in $\mathbf{v}$ and $\mathbf{j}$ and their derivatives, thereby linearizing all the equations.

2. We assume electrical neutrality, with $n_i Z$ equal to n_e. In those situations where an electrical field may exist in a plasma deviations from electrical neutrality must be considered in Poisson's law, equation (2-16) below, but may be ignored in the dynamical equations for $\mathbf{j}$ and $\mathbf{v}$.

3. We substitute a scalar pressure for the stress tensor, in accordance with equation (2-7).

Addition of $n_i m_i \partial v_i/\partial t$ and $n_e m_e \partial v_e/\partial t$ now gives the familiar linearized equation of motion

$$\rho \frac{\partial \mathbf{v}}{\partial t} = \mathbf{j} \times \mathbf{B} - \nabla p - \rho \nabla \phi \tag{2-11}$$

The interaction terms P_{ei} and P_{ie} have cancelled out, by Newton's third law of motion. Subtracting $n_e \partial \mathbf{v}_e/\partial t$ from $n_i Z \partial \mathbf{v}_i/\partial t$ yields

$$\frac{m_i m_e c^2}{Z\rho e^2}\frac{\partial \mathbf{j}}{\partial t} = \mathbf{E} + \mathbf{v} \times \mathbf{B} - \eta j + \frac{c}{eZ\rho}\left(m_i \nabla p_e - Zm_e \nabla p_i - (m_i - Zm_e)\mathbf{j} \times \mathbf{B}\right) \quad (2\text{-}12)$$

We see that the gravitational force has cancelled out.

In deriving equation (2-12) we have assumed

$$\mathbf{P}_{ei} = \frac{\eta e n_e}{c}\mathbf{j} \quad (2\text{-}13)$$

where η is a suitable constant of proportionality. It is reasonable to assume that the momentum exchanged between positive ions and electrons should be proportional to the relative velocity of the two types of particles. Because of the velocity dependence of the collisional cross section, there will also be contributions to $\mathbf{P}_{ei}$ proportional to ∇T and to $\mathbf{B} \times \nabla T$; these thermoelectric terms are evaluated in Chapter 5, but are omitted here from equation (2-12), since the influence of thermoelectric effects on the dynamics of a plasma has yet to be considered in detail.

When $\partial\mathbf{j}/\partial t$, $\mathbf{B}$, ∇p_e and ∇p_i all vanish, equation (2-12) reduces to Ohm's Law, with η equal to the electrical resistivity. We may therefore refer to equation (2-12) as the "generalized Ohm's Law." As we shall see in Section 5.4, η is about twice as great for currents perpendicular to a strong magnetic field as for currents parallel to $\mathbf{B}$ (or in the absence of $\mathbf{B}$). Hence the $\eta\mathbf{j}$ term should strictly be resolved into its components, with $\eta_\perp j_\perp$ and $\eta_\parallel j_\parallel$ perpendicular and parallel, respectively, to $\mathbf{B}$. To a first approximation, however, an isotropic η may be assumed. When more than one type of positive ion is present, the generalized Ohm's Law becomes more complicated, and another equation must be added to determine the relative velocities of the two types of ions.

Equations (2-11) and (2-12) must be supplemented by the equations of continuity of matter and electric charge. From the general equation of continuity obtained in the Appendix, equation (6-8), we may write

$$\frac{\partial \rho}{\partial t} + \nabla \cdot (\rho \mathbf{v}) = 0 \tag{2-14}$$

$$\frac{\partial \sigma}{\partial t} + \nabla \cdot \mathbf{j} = 0 \tag{2-15}$$

where σ is the charge density in electromagnetic units. Separate equations for $\partial n_e/\partial t$ and $\partial n_i/\partial t$ may be obtained by substituting equations (2-8) and (2-9) in these results.

Finally we have Maxwell's equations for the electromagnetic field, completing the list of basic relations. These equations may be written, in e. m. u.,

$$\nabla \cdot \mathbf{E} = 4\pi c^2 \sigma = 4\pi ec(n_i Z - n_e) \tag{2-16}$$

$$\nabla \cdot \mathbf{B} = 0 \tag{2-17}$$

$$\nabla \times \mathbf{E} = -\frac{\partial \mathbf{B}}{\partial t} \tag{2-18}$$

$$\nabla \times \mathbf{B} = \frac{1}{c^2}\frac{\partial \mathbf{E}}{\partial t} + 4\pi \mathbf{j} \tag{2-19}$$

Equation (2-15) above may also be derived from equations (2-16) and (2-19).

The use of $\nabla \times \mathbf{B}$ instead of $\nabla \times \mathbf{H}$ in equation (2-19) may seem somewhat unusual. Since a plasma is a diamagnetic medium, it seems desirable to utilize **B** to emphasize that the actual magnetic field inside the gas may differ from the field produced by external currents. However, since the permeability is not a very useful concept for a plasma it seems best to treat all plasma currents explicitly in equation (2-19). Once this is done all formal distinction between **B** and **H** vanishes, and if **B** is used, $\nabla \times \mathbf{B}$ is the proper quantity in equation (2-19). Anyone who so wishes may substitute **H** for **B** throughout this volume; if **H** is used, the oersted should replace the gauss as the unit of measurement.

To these equations we must add another relationship determining the temperature and hence the pressure. In certain simple cases one may use the adiabatic relations discussed in Section 1.4. If collisions produce a nearly isotropic velocity distribution, but energy losses are negligible, then γ in equation (1-33) must equal 5/3. If collisions are negligible, but all changes are two-dimensional, transverse to **B**, $p_\perp$ replaces p in the equations above, with $T_\perp$ proportional to B as in equation (1-35). For slow changes in density the temperature is determined by the equation of energy balance, including such effects as resistance losses (ohmic or Joule heating), radiation and absorption of electromagnetic waves, and heat conduction.

In equation (2-12) terms of order m_e/m_i have been retained. These terms are required, for example, to give correct results for low-frequency waves in plasmas of very low density. Under most conditions, however, these terms are negligible. If we ignore not only terms in m_e/m_i but also the terms in $\partial \mathbf{j}/\partial t$ and $\partial \mathbf{v}/\partial t$, considering changes so slow that inertial effects are negligible, we obtain the much simpler equations

$$\nabla p = \mathbf{j} \times \mathbf{B} - \rho \nabla \phi \tag{2-20}$$

$$\mathbf{E} + \mathbf{v} \times \mathbf{B} = \eta \mathbf{j} + \frac{c}{en_e}(\nabla p_i + \rho \nabla \phi) \tag{2-21}$$

where equation (2-11) has been used in eliminating $\mathbf{j} \times \mathbf{B}$ from equation (2-12). These equations form the basis for the analysis of equilibrium configurations in Chapter 4.

2.3 Relation between Macroscopic and Microscopic Velocities

Equations (2-20) and (2-21) yield the macroscopic values of $\mathbf{j}$ and $\mathbf{v}$ when conditions in the plasma are changing slowly. To solve for $\mathbf{j}$ and $\mathbf{v}$ we take the vector product of these equations with **B**, and obtain

$$\mathbf{j}_\perp = \frac{\mathbf{B} \times \nabla p}{B^2} \tag{2-22}$$

$$\mathbf{v}_\perp = \frac{\mathbf{B}}{B^2} \times \left(-\mathbf{E} + \frac{c}{en_e} \nabla p_i \right) \tag{2-23}$$

where we have assumed that both η and ϕ are zero. These equations must be satisfied in any quasi-steady solution. It is of interest to note that the roles of the two basic equations are reversed from the usual custom, since the equation of motion now determines the current density, while the generalized Ohm's law determines the velocity. This inversion is an important feature characterizing a plasma in a quasi-steady state in a magnetic field.

The motions determined from equations (2-22) and (2-23) do not agree with the microscopic particle drifts discussed in Section 1.2. It is only the effect of **E** which is the same in the two pictures; the macroscopic mass velocity resulting from **E** is the same in equation (1-9) and (2-23), and no current is produced, since electrons and positive ions drift at the same rate. On the other hand, the microscopic drifts produced by inhomogeneous magnetic fields do not appear in the macroscopic **v** and **j**, while the macroscopic velocities and currents associated with pressure gradients have apparently no counterpart in the single-particle drifts. Some confusion has arisen in the past in connection with this apparently paradoxical result. We analyze here the difference between these two types of mean velocity.

The drift velocity $\mathbf{w}_D$ is defined as the mean velocity of the guiding centers in a volume element. If we consider all the particles gyrating about these guiding centers, the phases of gyration of the different particles will be random, the velocities of gyration will average out, and the mean velocity of such particles, which we may denote by $\mathbf{v}'$, will equal $\mathbf{w}_D$. However, $\mathbf{v}'$ is not the macroscopic velocity $\mathbf{v}$, which is defined as the mean velocity of all the particles which are located in a volume element, regardless of where their guiding centers are located. Thus to obtain $\mathbf{v}$ we must correct $\mathbf{v}'$ by first omitting from the average those particles which are outside the volume element, although their guiding centers are inside. Then we must include in the average those particles which are inside the volume element at a particular moment, but whose guiding centers are outside. When this correction is carried out, it is found that the computed value of $\mathbf{v}$ agrees with equation (2-23). Analyses along this line have been given by Schlüter (11) and Spitzer (12).

As a substitute for the detailed analysis, a more general physical explanation of the results will be given here for a number of specific cases. When a pressure gradient is present, the macroscopic equations yield both a current density $\mathbf{j}$ and a velocity $\mathbf{v}$. On the other hand, the particle drift velocity $\mathbf{w}_D$ is zero, provided that $\mathbf{B}$ is uniform. Actually we shall see below—equation (4-1)—that a pressure gradient is accompanied by a gradient of B, but the $\mathbf{w}_D$ due to this field gradient becomes negligibly small compared to the corresponding macroscopic velocity, found from equation (2-25), as p becomes small compared to $B^2/8\pi$. To simplify the problem we shall assume that B is uniform and w_D zero.

Figure 2.1 illustrates how a pressure gradient can produce macroscopic currents and velocities across the lines of force, even though the guiding centers of the individual particles have no transverse drifts. The orbits of those particles are shown whose guiding centers lie in the plane $A'A$, perpendicular to the plane of the diagram; the magnetic field is also perpendicular to the plane of the diagram. The number of parti-

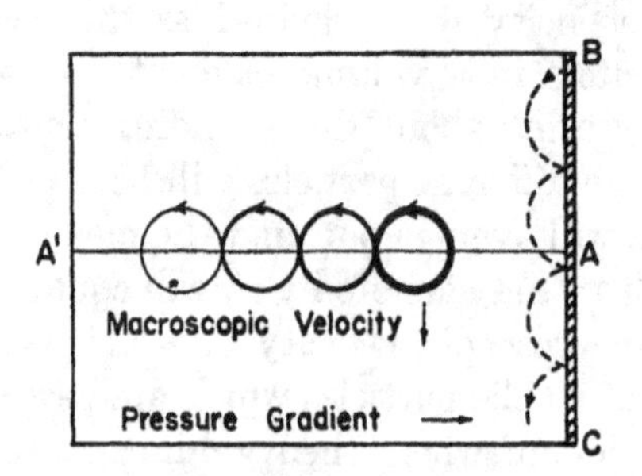

Figure 2.1. Macroscopic velocity resulting from a pressure gradient. The thickness of each line is proportional to the number of particles gyrating in that circle.

cles gyrating in each circular orbit is represented by the width of the line; thus there are more particles gyrating on the right-hand side of the figure than on the left. It is clear that through a small element in the plane $A'A$ more particles will move downward than upward. The quantitative analysis shows that both positive ions and electrons contribute to the current density, each proportionally to the pressure gradient for that particle. The macroscopic velocity depends only on the positive ions, owing to their much greater mass. Hence the electron pressure gradient does not appear in equation (2-23).

It is clear, of course, that the total number of particles crossing the plane $A'A$ must be the same on either the microscopic or the macroscopic picture. End effects must be taken into account to obtain agreement. We assume that the plasma is bounded, on the right-hand side, by the infinite plane BAC. The particles which are reflected from this plane will travel downwards, as shown by the dashed line in Figure 2.1. When these downward-moving particles are considered, the total flux of particles across the plane $A'A$ is the same, whether determined microscopically or macroscopically.

The presence of particles reflected from the wall is necessary to reconcile the microscopic and macroscopic pictures. The importance of such particles was originally pointed out by Bohr and van Leeuwen, who explained why an isothermal electron gas of uniform density, confined by reflecting walls, should not be diamagnetic in the presence of a magnetic field. The diamagnetism resulting from all the freely gyrating electrons is exactly cancelled by the wall current, consisting of electrons reflected from the wall, as in Figure 2.1. When the electron pressure is not uniform, macroscopic currents appear, and the gas becomes diamagnetic as shown in Section 4.2 below.

We have seen that currents and macroscopic velocities can be present even when the individual particles show no drift. Conversely, we shall now see that the particle drifts caused by inhomogeneities in magnetic field strengths do not produce any macroscopic effects. Equations (2-22) and (2-23) do not give any systematic currents or velocities when **B** is inhomogeneous, with **E** and ∇p zero. This result must be reconciled with equations (1-11) and (1-12) of the previous chapter.

To understand the reason for the apparent paradox, let us again consider a region enclosed by perfectly reflecting walls. Let there be an arbitrary field **B** inside this region, such that the force on a charged particle is $q\mathbf{w} \times \mathbf{B}$. Let us now imagine that at some initial time t_0 a group of identical particles is placed within the region. (Particles of opposite sign must also be included to preserve electrical neutrality, but we may ignore these, since collisions will be neglected.) We assume that all these particles have a scalar velocity within a narrow range Δw, that the velocity distribution is isotropic, and that the number of particles per unit volume is initially uniform. In other words, the density of particles in phase space is uniform throughout the region, for w within the range Δw, but vanishes for scalar velocities outside this range. Liouville's theorem, which is applicable here, states that the density in phase space is constant along a dynamical trajectory. Since the assumed force is perpendicular to **w**, the direction of particle motion will change along a dynamical trajectory, but the scalar velocity will not. Since the initial density in phase space is independent of the direction of motion and also independent of position in space, it follows that the motions of the particles will not produce any change whatever in the phase-space density of particles. Hence the initial distribution will remain constant in time. The macroscopic current and velocity obviously vanish, since the velocity distribution is isotropic. Consideration of particles with other velocities will not affect this result. We conclude that in a region where the velocity distribution is initially isotropic and the density and pressure are uniform, no macroscopic velocities or currents can appear, regardless of what magnetic field, **B**, may be present, provided that $\partial\mathbf{B}/\partial t$ vanishes.

As before, the total flux of particles across any infinite plane must be the same, whether computed macroscopically or microscopically. The flux of wall-reflected particles must therefore exactly cancel the net number of particles drifting across the plane because of magnetic inhomogeneities.

When collisions are important, this proof may be modified. Let us first consider a system without collisions, but with a velocity distribution which is everywhere isotropic and Maxwellian, with a constant kinetic temperature over the region considered. It is clear from the above arguments that such a distribution is self-maintaining and that macroscopic velocities and currents vanish. The introduction of collisions will clearly not alter the velocity distribution. Evidently in a closed system in thermodynamic equilibrium $\mathbf{j}$ must vanish, a conclusion emphasized by Cowling (4).

2.4 Electric Currents

The ratio of $\mathbf{j}$ to $\mathbf{E}$ in specific situations is frequently called the conductivity. When Ohm's law is valid this ratio is indeed the normal conductivity, or $1/\eta$. In other situations, however, $\mathbf{j}$ and $\mathbf{E}$ are not necessarily parallel, and the ratios of their components in different directions depend on the detailed solution of the basic equations. In fact, the ratio j_x/E_x may lie anywhere from zero to infinity, depending on the previous history of the plasma. There has been some confusion on this subject in the past, and it is worth while to discuss in some detail the magnitude of $\mathbf{j}$ under different conditions.

There are circumstances in which Ohm's law in the simple form is always obeyed in a plasma. For example, if the plasma is in a steady state, with $\mathbf{v}$ everywhere zero, and if the gravitational potential vanishes, equation (2-21) gives

$$\eta \mathbf{j} = \mathbf{E} - \frac{c}{en_e} \nabla p_i \tag{2-24}$$

According to equation (2-20), $\mathbf{j}$ can have no component parallel to ∇p, if a magnetic field is assumed to be present. If ∇p_i and ∇p_e are assumed everywhere parallel, then $\mathbf{j}$ can have no component parallel to ∇p_i, and according to equation (2-24) $\eta\mathbf{j}$ must equal the component of **E** parallel to the current. Thus in a certain sense Ohm's law in the simple form is obeyed. The remaining component of **E** balances out the gradient of p_i to produce zero velocity in equation (2-23). As Schlüter (10) has shown, a plasma in a magnetic field has a remarkable tendency to approach the equilibrium situations in which equation (2-24) is obeyed.

If we take the scalar product of equation (2-24) with **j**, in this idealized case, we obtain

$$\eta j^2 = \mathbf{j} \cdot \mathbf{E} \tag{2-25}$$

Evidently $\mathbf{j} \cdot \mathbf{E}$ is the rate at which work is done on the electric current per unit volume, while ηj^2 represents the heating of the gas by the current. To maintain a steady state the heat generated must be radiated away as fast as it is produced.

In other situations, $\eta\mathbf{j}$ and **E** may have little relationship to each other, although the heat dissipated because of the resistance losses is always ηj^2. To consider the simplest possible case that illustrates the essential features of the problem, let us suppose that an electric field **E**, transverse to the magnetic field, is applied abruptly, increasing from zero to a constant value at the time $t = 0$, when $\mathbf{v} = \mathbf{j} = 0$. We consider an infinite medium, with no gravitational field, and let the pressures be everywhere constant. We shall neglect all induced electric and magnetic fields produced by the changing current, although in fact these may prevent penetration of the electrical field into the plasma (skin effect). If now we express η in terms of the collision frequency ν, by means of equation (5-32), and use equation (2-11) to eliminate **v** from equation (2-12), assuming that $\mathbf{j} \cdot \mathbf{B}$ vanishes and neglecting m_e/m_i compared to unity, we find, after some rearrangement

$$\frac{\partial \mathbf{j}}{\partial t} = \frac{\nu \mathbf{E}}{\eta} - \omega_{ci}\omega_{ce} \int_0^t \mathbf{j}\,dt - \omega_{ce}\frac{\mathbf{j} \times \mathbf{B}}{B} - \nu \mathbf{j} \tag{2-26}$$

where ω_{ce} and ω_{ci} are the cyclotron frequencies of electrons and positive ions, respectively. We investigate the solution of equation (2-26) in various simplified situations.

a. No Magnetic Field, $\omega_{ci} = \omega_{ce} = 0$. In this case we have the simple solution

$$\mathbf{j} = \frac{\mathbf{E}}{\eta}(1 - e^{-\nu t}) \tag{2-27}$$

Evidently $\mathbf{j}$ approaches its final value very closely within a few collision times.

b. $\omega_{ci} = 0$, ω_{ce} *Finite.* Since we are interested primarily in the final steady current, we set $\partial \mathbf{j}/\partial t$ equal to zero. If we take the z axis in the direction of $\mathbf{B}$, and the y axis along $\mathbf{E}$, equation (2-26) now yields the solution

$$j_y = \frac{E_y}{\eta} \times \frac{1}{1 + (\omega_{ce}/\nu)^2} \tag{2-28}$$

$$j_x = \frac{-E_y}{\eta} \frac{\omega_{ce}/\nu}{1 + (\omega_{ce}/\nu)^2} \tag{2-29}$$

If the term $\partial \mathbf{j}/\partial t$ is retained, j_y and j_x oscillate around their final values with an angular frequency ω_{ce}, the oscillations damping out in a few collision times. Thus equations (2-28) and (2-29) are valid only for times large compared to $1/\nu$.

When ω_{ce} much exceeds ν the current parallel to the applied electrical field is much reduced, and varies linearly with the collision frequency (since ν and η are proportional). This result may be understood simply on the basis of the single-particle drifts. When an electric field E_y is applied to a group of electrons, transverse to a magnetic field in the z direction, the electrons start to move in the y direction, but after a net displacement d_y in this direction they end up drifting in the x direction with the velocity E_y/B_z. After the electrons have collided with stationary positive ions and lost their momentum in the x direction, they suffer another displacement d_y, on the average, before drifting off in the x direction. The more frequent the collisions, the more frequent the displacements in the y

direction and the greater the current in this direction. The current in the x direction, called the Hall current, is independent of ν, if ω_{ce} much exceeds ν, and is simply equal to the charge density $-n_e e/c$ times the transverse drift velocity E_y/B.

c. ω_{ci}, ω_{ce} Both Finite. While the detailed solution of equation (2-26) is somewhat involved in this general case, the physical results are simple and physically understandable. In a time comparable with $1/\omega_{ci}$ the positive ions are also influenced by the electric field and begin to drift across the magnetic field with the velocity E_y/B. Their contribution to j_y (the current density parallel to **E**) far outweighs that of the electrons, if ω_{ce}/ν is large, while their contribution to j_x (transverse to **E** and **B**) tends to cancel the electron Hall current. In the steady state both electrons and positive ions drift in the x direction with a velocity E_y/B. When electrons collide with the positive ions they do not lose any momentum in the x direction and hence do not suffer any displacement in the y direction, and therefore the electron current parallel to **E** vanishes. For the same reason, the positive-ion current parallel to **E** vanishes. The fact is, of course, that all currents in the final state must vanish, since to an observer traveling with the drift velocity no electric field is present. More generally, equation (2-22) shows that steady-state electric currents transverse to **B** are produced only by pressure gradients transverse to the magnetic field.

A detailed solution of equation (2-26), retaining all terms, but with ν assumed small compared to ω_{ce}, shows that the contribution of positive ions to j_y equals the electron contribution, given by equation (2-28), when t equals $\nu/\omega_{ci}\omega_{ce}$. For greater values of the time the positive-ion current continues to rise for a while, but then oscillates with an angular frequency ω_{ci}, the oscillations dying out as exp $(-\nu t\omega_{ci}/\omega_{ce})$. We see that equation (2-28) is valid only for t less than $\nu/\omega_{ci}\omega_{ce}$. However it has already been shown that equations (2-28) and (2-29) are correct only if t exceeds $1/\nu$, since for shorter times the neglected inertial terms, proportional to $\partial \mathbf{j}/\partial t$, become important. Evidently in an ionized gas these equations are valid only in the special case where ν exceeds $(\omega_{ce}\omega_{ci})^{\frac{1}{2}}$, and then only for a particular range of values for t. In a weakly ionized gas much more time is required for all the gas to attain the drift velocity E_y/B, and the range of validity of these equations is much increased.

d. Polarization. While the detailed general solution of equation (2-26) is too complicated to warrant further discussion here, one simple result follows immediately from this equation. We have seen that in the steady state which is finally reached after the imposition of a constant **E**, **j** must vanish. Hence we obtain immediately the following condition on the integral of **j** over time

$$\int_0^\infty \mathbf{j}dt = \frac{\nu}{\omega_{ci}\omega_{ce}\eta}\mathbf{E} = \frac{\rho}{B^2}\mathbf{E} \tag{2-30}$$

where ρ is again the mass density and where we have again used equation (5-32) for ν/η. Essentially the total current flow is limited by the condition that the force $\mathbf{j} \times \mathbf{B}$ accelerates the gas up to the drift velocity E/B. Equation (2-30) also indicates that for sufficiently large t, the time integral of the Hall current must vanish exactly.

Equation (2-30) is derived for the situation where **E** is constant after an abrupt initial increase. If **E** changes slowly, so that its relative change in the time $1/\omega_{ci}$ is small, then **v** is closely equal to $\mathbf{E} \times \mathbf{B}/B^2$ at all times, and from the equation of motion, equation (2-11), we find

$$\mathbf{j}_\perp = \frac{\rho}{B^2}\mathbf{B} \times \frac{d\mathbf{v}}{dt} = \frac{\rho}{B^2}\frac{d\mathbf{E}_\perp}{dt} \tag{2-31}$$

This is the equation for a polarizable medium, with a polarizability ρ/B^2. The dielectric constant, K, then becomes

$$K = 1 + \frac{4\pi\rho c^2}{B^2} \tag{2-32}$$

Thus a fully ionized gas in a magnetic field, while it does not show a simple conductivity, behaves like a dielectric for the component of **E** perpendicular to **B**. If ρ and **B** are time dependent, the polarization cannot, in general, be found by use of the dielectric constant K given in equation (2-32), but must be found directly from the basic equations.

2.5 Motion of Material across Lines of Force

As pointed out by Alfvén (1), the lines of force within a perfectly conducting gas tend to be "frozen in" the material. This concept may appear to be somewhat vague, since electromagnetic theory offers no unique definition of the motion of a line of force. To make this idea more precise, we may say that in a conducting gas the magnetic flux Φ through any closed contour, each element of which moves with the local gas velocity $\mathbf{v}$, tends to remain constant. If Φ through every contour is strictly constant during the motion, Newcomb (8) has shown that the magnetic lines of force can always be taken to be moving entities, each element of which moves with the local velocity $\mathbf{v}$. This representation is not always unique, and thus some magnetic fields can be represented by lines which do not move with the local fluid velocity; for example, an arbitrary rotation about an axis of symmetry can be assumed. If Φ is not constant through a contour following the fluid, but $\mathbf{E}\cdot\mathbf{B}$ vanishes, $\mathbf{B}$ can still be represented by lines of force, but these are not fixed with relation to the fluid. If $\mathbf{E}\cdot\mathbf{B}$ differs from zero, the lines of force do not necessarily retain their identity as $\mathbf{B}$ changes, and representation of the field with lines of force moving at any velocity may not be possible.

To determine the conditions under which the lines of force may be regarded as frozen in the fluid, we shall compute the general conditions under which the magnetic flux is constant through a closed contour moving with the fluid. The velocity at which a plasma diffuses across a strong magnetic field as a result of particle collisions and other effects will then be discussed.

a. Change of Flux through a Moving Surface. The change of Φ through a moving surface results from two causes: first, the change of **B** with time at various points on the surface, with **v** set equal to zero, and secondly the motion of the contour itself, which may result in encompassing more or less flux. The change of field with time produces a change of Φ given by

$$\frac{\partial \Phi}{\partial t} = \iint \frac{\partial \mathbf{B}}{\partial t} \cdot \mathbf{dS} \tag{2-33}$$

where **dS** is an element of area. The change of Φ resulting from the motion alone is found from the rate at which the contour moves across the lines of force, with the time derivative of **B** ignored and the lines of force motionless. If **ds** is an element of length of the contour, then $\mathbf{v} \times \mathbf{ds}$ is the area swept over by the element, per unit time; the flux through this area is $\mathbf{B} \cdot \mathbf{v} \times \mathbf{ds}$. Hence the change of Φ resulting from the motion alone is

$$\frac{\partial \Phi}{\partial t} = \int \mathbf{B} \cdot \mathbf{v} \times \mathbf{ds} \tag{2-34}$$

integrated around the contour.

Equations (2-33) and (2-34) may now be combined to yield the total time derivative of Φ. In equation (2-33) we substitute from equation (2-18), while in equation (2-34) we first interchange the dot and vector products, and then transform the line integral to a surface integral, by means of Stokes' theorem. We obtain

$$\frac{d\Phi}{dt} = -\iint \nabla \times \{\mathbf{E} + \mathbf{v} \times \mathbf{B}\} \cdot d\mathbf{S} \tag{2-35}$$

The condition that $d\Phi/dt$ vanish for any surface implies that the integrand of equation (2-35) must vanish everywhere. Hence

$$\nabla \times \{\mathbf{E} + \mathbf{v} \times \mathbf{B}\} = 0 \tag{2-36}$$

Thus, if equation (2-36) is satisfied the magnetic flux will be invariant through any surface moving with the velocity **v**.

To investigate whether equation (2-36) is satisfied we may take the curl of the generalized Ohm's law (2-12). Evidently if a number of subsidiary conditions are satisfied, equation (2-36) will be fulfilled in a fully ionized gas. If we eliminate the $\mathbf{j} \times \mathbf{B}$ term by means of equation (2-11) these conditions are that $\nabla \times (\partial \mathbf{j}/\rho \partial t)$ and $\nabla \times (\partial \mathbf{v}/\partial t)$ are negligible, that p_i and p_e are functions of ρ alone, and that η is negligibly small. For relatively slow motions of a highly conducting gas these conditions are usually satisfied approximately, but in the general case the motion of the gas cannot necessarily be identified precisely with any motion assignable to the lines of force.

The natural decay of a magnetic field resulting from ohmic losses is readily found from equation (2-35) if the $\eta \mathbf{j}$ term is retained in equation (2-12), but the terms in $\partial \mathbf{j}/\partial t$, $\partial \mathbf{v}/\partial t$, ∇p_i, ∇p_e, and $\nabla \phi$ are ignored. Eliminating $\mathbf{j}$ by means of equation (2-19) (with $\partial \mathbf{E}/\partial t$ neglected), and assuming η constant, we obtain

$$\frac{\partial \Phi}{\partial t} = \frac{\eta}{4\pi} \iint \nabla^2 \mathbf{B} \cdot d\mathbf{S} \tag{2-37}$$

If we approximate $\nabla^2 \mathbf{B}$ by $\mathbf{B}/L^2$, where L is a length characteristic of the system, we see that Φ decays exponentially with a time constant, τ, given by

$$\tau \approx \frac{4\pi L^2}{\eta} \approx 2 \times 10^{-13} T^{3/2} L^2 \text{ sec} \tag{2-38}$$

where the numerical value of η for an electron-proton gas has been inserted from equation (5-37), with the quantity $\ln \Lambda$ set equal to ten. The time τ is simply the length of time required for the ohmic losses ηj^2 to dissipate an energy comparable with the magnetic energy density $B^2/8\pi$.

b. Diffusion across a Strong Magnetic Field. Let us now consider a much more special case, where a plasma is confined by a strong magnetic field, $\mathbf{B}$, whose value at each point remains constant in time, independently of the behavior of the plasma. This condition on $\mathbf{B}$ can readily be satisfied if the magnetic field within the region considered is produced by currents which are outside the region and constant in time;

this condition requires that the gas pressure, p, within the region is negligibly small compared to $B^2/8\pi$, so that the plasma currents required by equation (2-22) for a quasi-steady state produce a negligible effect on $\mathbf{B}$.

The diffusion velocity across the magnetic field, resulting from collisions of electrons with positive ions, may now be obtained from equations (2-20) and (2-21) for a quasi-steady state. When finite resistivity is taken into account, a term $-\eta\mathbf{j} \times \mathbf{B}/B^2$ must be added to the right-hand side of equation (2-23). If we substitute for $\mathbf{j} \times \mathbf{B}$ from equation (2-20) we obtain an additional contribution to $\mathbf{v}_\perp$, which we denote by $\mathbf{v}_{D_\eta}$ to indicate the transverse diffusion velocity resulting from finite resistivity. We find

$$\mathbf{v}_{D_\eta} = -\frac{\eta}{B^2}\nabla p = -\frac{1.78 \times 10^{-3} Z \ln \Lambda}{B^2 T^{1/2}} \nabla n \qquad (2\text{-}39)$$

where we have utilized equation (5-42) for the transverse η in an ionized gas, and have assumed constant T. It should be noted that the component of $\mathbf{v}_\perp$ perpendicular to ∇p_i, given in equation (2-23), does not impair the confinement of the plasma, since the material motion is parallel to the isobars and shows no divergence. If ∇T is not zero, the thermoelectric effects discussed in Section 5.5 should be taken into account.

If we regard equation (2-39) as the velocity of the fluid relative to the magnetic lines of force, this result is quite general, restricted only by the neglect of the inertial terms and the gravitational potential in the general macroscopic equations. Equations (2-38) and (2-39) provide different examples of the same basic process. In fact, equation (2-37) may be derived from the condition that Φ changes as a result of relative motion between fluid and lines of force with the velocity v_{D_η} given in equation (2-39). This derivation requires the restriction that $\mathbf{j}$ be perpendicular to $\mathbf{B}$; if this condition is satisfied, $\mathbf{E}\cdot\mathbf{B}$ will vanish as the field changes, and Newcomb (8) has shown that the lines of force will retain their identity during this decay.

The outwards motion of a confined plasma across a strong magnetic field is referred to as collisional diffusion. Measures on a cesium plasma by D'Angelo and Rynn (5) indicate a diffusion rate in general agreement with equation (2-39) for collisional diffusion.

When positive ions of different types are present, collisions again produce systematic motions across the line of force, except that now the different ions can move in different directions. The diffusion velocities in this case, which have been computed by Spitzer (12), tend toward an equilibrium in which $\nabla p_i/n_i$ is proportional to Z_i for each type of ion; thus the more highly charged particles tend to be concentrated in the denser regions of the plasma.

Encounters between identical particles do not produce any appreciable diffusion, at least in the first approximation. To understand this rather surprising result we turn to the microscopic picture to see how, in at least one special case, the diffusion produced by such collisions is negligibly small compared to the diffusion found from equation (2-39), associated with collisions between unlike particles. Specifically, the motion of the center of gravity of the guiding centers will now be considered. Let $\mathbf{r}$ be the position of a particle in some coordinate system, and let $\mathbf{r}_c$ be the position of the particle's guiding center. Then we have

$$\mathbf{r}_c = \mathbf{r} + \frac{m}{qB^2}\mathbf{w} \times \mathbf{B} \tag{2-40}$$

Consider now an assembly of N identical particles. The center of gravity of the guiding centers of these particles is located at the position $\Sigma \mathbf{r}_c/N$, where the summation extends over all the particles. The motion of this center of gravity with time is given by

$$\frac{1}{N}\frac{d}{dt}\sum \mathbf{r}_c = \frac{1}{N}\sum \mathbf{w} + \frac{m}{NqB^2}\sum \frac{d\mathbf{w}}{dt} \times \mathbf{B} \tag{2-41}$$

provided we assume that $\mathbf{B}$ is uniform through space. If now we take equation (1-1) for $d\mathbf{w}/dt$, with $\mathbf{E}$ set equal to zero, we find that the two terms on the right-hand side of equation (2-41) cancel out, provided that there is no mean velocity parallel to **B**. A term representing the forces of mutual interaction between particles must be added to equation (1-1), but thanks to Newton's third law this term cancels out on summation over all particles, even if the forces are long range. On this microscopic picture there is detailed balancing in the collisional drifts of the guiding centers. An encounter between two identical particles shifts the two guiding centers by equal and opposite amounts. On the other hand, when two particles of opposite signs collide, their guiding centers move in the same direction, and a substantial shift in the center of gravity is produced thereby.

As we shall now see for one special case, a fixed center of gravity implies a very marked restriction on the net diffusion. Let a plasma be confined between two infinite parallel planes at $x = x_1$ and $x = x_2$, and let a strong uniform magnetic field extend in the z direction. The two planes will be assumed perfectly reflecting. At each reflection of a particle, the guiding center will shift abruptly in the y direction, parallel to the plane, with no motion in the x direction. Thus reflections from the walls will not change the x component of the center of gravity of all the particles. If now the gas is initially near one of the walls, with a mean x_c close to x_1 or to x_2, the gas will remain near that wall indefinitely. Hence in this sense no diffusion results, on the average, although collisions must affect the detailed spatial distribution of the plasma.

A more detailed examination of the transverse diffusion associated with ion-ion encounters shows that macroscopically this diffusion results from the viscous drag produced by such encounters. We evaluate the drift velocity corresponding to this diffusion in the simple case where a shearing velocity, $\mathbf{v}$, is present transverse to **B**; we assume that the motion is steady, with $\partial\rho/\partial t$ equal to zero, and that $\mathbf{v}$ is parallel to the surfaces of constant density and temperature. Under these conditions $\nabla\cdot\mathbf{v}$ and $\mathbf{v}\cdot\nabla\mu$ vanish, where μ is the coefficient of viscosity; the viscous force per unit volume which appears in the general Navier-Stokes equations, and which must be added on the right-hand side of equation (2-11), reduces to $\nabla \cdot (\mu \nabla \mathbf{v})$. In a steady state this viscous drag parallel to $\mathbf{v}$ must be balanced by a force $\mathbf{j} \times \mathbf{B}$, if the gravitational force is neglected; by assumption the pressure gradient has no component in this direction. Since collisions among electrons do not contribute appreciably to the viscosity, this drag is exerted primarily on the positive ions and the resulting current will be carried by the ions. Hence $\mathbf{j}$ equals $n_i Ze\mathbf{v}_{D\mu}/c$, where $\mathbf{v}_{D\mu}$ is the resulting ion drift, transverse to the magnetic field. The generalized equation (2-11) then yields

$$\mathbf{v}_{D\mu} = -\frac{c}{en_e B^2} \mathbf{B} \times (\nabla \cdot \mu \nabla \mathbf{v}) \tag{2-42}$$

In the general case the value of μ is different for different components of the stress tensor. Here we consider the special case where $\mathbf{v}$ is not only transverse to $\mathbf{B}$ but also has no gradient in the direction of $\mathbf{B}$. The shearing stress will then lie in the plane transverse to $\mathbf{B}$, and we may replace μ by $\mu_\perp$, whose value in a strong magnetic field is given in equation (5-55) below.

If we assume that $\mathbf{E}$ vanishes, $\mathbf{v} \times \mathbf{B}$ in the steady state may be expressed in terms of p_i from equation (2-21), neglecting the small $\eta\mathbf{j}$ term. If we further idealize the problem by assuming that $\mathbf{B}$ has the same constant value everywhere, we find

$$\mathbf{v}_{D\mu} = \frac{c^2}{e^2 n_e B^2} (\nabla \cdot \mu_\perp \nabla) \left(\frac{\nabla p_i}{n_e} \right) \tag{2-43}$$

The assumption that $\mathbf{E}$ vanishes may not be realistic; in particular, the current associated with $\mathbf{v}_{D\mu}$ will tend to set up electric fields that reduce the shearing velocity to zero. If equation (2-43) is valid, it is readily shown that in the one-dimensional isothermal case $v_{D\mu}$ is proportional to $n\partial/\partial x(\partial^2 n/n\partial x^2)$, a result that may be compared with equation (2-40). Evidently in this higher order diffusion process $v_{D\mu}$ vanishes if n varies exponentially with x.

A variety of observations indicate that in the presence of

fluctuating or turbulent fields the diffusion of plasma across the lines of force can be substantially enhanced above the collisional diffusion rate predicted by equation (2-39). Bohm (2) and his collaborators postulate that this diffusion is caused by irregular fluctuations in the gas, akin to the turbulence observed in ordinary liquids. If we assume that this process is unrelated to collisions between particles the magnitude of the diffusion velocity to be expected may be computed from the following dimensional argument.

Let us assume that the particle flux is given by a diffusion coefficient D times the density gradient, ∇n. Then

$$n\mathbf{v}_D = D\nabla n \tag{2-44}$$

and D has the dimensions of a velocity times a length. In ordinary diffusion of neutral atoms through a gas D is about equal to λw_T, where λ is the mean free path and w_T is the thermal velocity. The diffusion velocity across a strong magnetic field, given in equation (2-39), corresponds to D equal to $\lambda w_T(a/\lambda)^2$, where a is the radius of gyration; this result may be verified directly with the aid of equation (5-32). Evidently the simplest form for D which is not related to λ is aw_T. If we substitute from equation (1-4) for a, D is about equal to ckT/eB. Bohm suggests dividing this result for D by another factor of 16. If we denote by v_{Dt} the diffusion velocity resulting from plasma turbulence, Bohm's hypothesis yields

$$v_{Dt} = -\frac{ckT}{16\,en_eB}\nabla n_e = -\frac{5.4 \times 10^2 T}{n_eB}\nabla n_e \tag{2-45}$$

The factor 16 in the denominator does not seem to have any particular theoretical justification, since no detailed derivation of this equation, in terms of a specific mechanism, has yet appeared. The dimensional derivation is not very conclusive, since D might well be multiplied by one of the several dimensionless parameters characterizing the plasma, such as m_i/m_e or the dielectric constant K, for example.

Experiments have not yet given a definitive verification of

equation (2-45). Indirect determinations by Stodiek, Ellis, and Gorman (13) on plasmas through which a substantial electric current was passing give v_D varying as T/B, but with a numerical value greater than the Bohm formula by about a factor three. On the other hand, measures by Post and his collaborators (9) on plasmas confined in a magnetic mirror, with no appreciable plasma current flowing, indicate diffusion velocities orders of magnitude less than those predicted from the Bohm formula. Evidently further research is required to indicate not only the conditions under which plasma fluctuations can increase the transverse diffusion but also the other detailed effects produced by such turbulent variations.

References

1. Alfvén, H., *Ark. Mat., Astr. Fysik*, **29B**, No. 2 (1942).
2. Bohm, D., *The Characteristics of Electrical Discharges in Magnetic Fields*, edited by A. Guthrie and R. K. Wakerling, McGraw-Hill, New York, 1949, Chapter 2, Section 5.
3. Chew, G. F., M. L. Goldberger, and F. E. Low, *Proc. Roy. Soc. (London)*, **236A**, 112 (1956).
4. Cowling, T. G., *Monthly Notices, Roy. Astron. Soc. (London)*, **92**, 407 (1932).
5. D'Angelo, N., and N. Rynn, *Phys. Fluids*, **4**, 1303 (1961).
6. Langmuir, I., *Phys. Rev.*, **33**, 954 (1929).
7. Lüst, R., *Fortschr. Physik*, **7**, 503 (1959).
8. Newcomb, W. A., *Ann. Phys.*, **3**, 347 (1958).
9. Post, R. F., R. E. Ellis, F. C. Ford, and M. N. Rosenbluth, *Phys. Rev. Letters*, **4**, 166 (1960).
10. Schlüter, A., *Z. Naturforsch.*, **5a**, 72 (1950).
11. Schlüter, A., *Ann. Physik*, **6–10**, 422 (1952).
12. Spitzer, L., *Astrophys. J.*, **116**, 299 (1952).
13. Stodiek, W., R. A. Ellis, and J. G. Gorman, *Nuclear Fusion Supplement*, Part I, 193 (1962).
14. Tonks, L., and I. Langmuir, *Phys. Rev.*, **34**, 876 (1929).
15. Watson, K. M., *Phys. Rev.*, **102**, 12 (1956).

CHAPTER 3

Waves in a Plasma

An ionized gas is capable of a wide variety of oscillatory motions. In general these oscillations may be exceedingly complex. We shall consider here infinitesimal disturbances in homogeneous media under relatively simple conditions. Considerable theoretical study has been given to three particular types of waves in a plasma: electromagnetic waves, hydromagnetic waves, and electrostatic waves. Although it seems unlikely that any of these idealized waves in pure form will be found in nature, except under controlled laboratory conditions, an understanding of these simple oscillations will give insight into the more complicated phenomena that may occur.

The analysis of these various types of waves is based primarily on Maxwell's equations, given in Chapter 2. If we take the curl of equation (2-18) and differentiate equation (2-19) with respect to time, eliminating $\nabla \times \partial B/\partial t$, we obtain the basic equation for electromagnetic waves,

$$\nabla^2 \mathbf{E} - 4\pi c^2 \nabla \sigma = \frac{1}{c^2} \frac{\partial^2 \mathbf{E}}{\partial t^2} + 4\pi \frac{\partial \mathbf{j}}{\partial t} \tag{3-1}$$

In deriving this result we have expanded the triple vector product and used equation (2-16) to eliminate $\nabla \cdot \mathbf{E}$. For E_κ, the component of $\mathbf{E}$ parallel to the direction of propagation, equation (3-1) is not useful, since for this component each side of the equation is identically zero. We use instead the relation

$$\frac{1}{c^2} \frac{\partial E_\kappa}{\partial t} + 4\pi j_\kappa = 0, \tag{3-2}$$

which follows at once from equation (2-19), since the derivative of $\mathbf{B}$ taken along the wave front vanishes.

To determine **E** equations (3-1) and (3-2) must be solved, together with the other equations in Chapter 2. The quantities σ and **j** are related by equation (2-15). The current density **j**, and the velocity **v** that may enter into the determination of **j**, are found from equations (2-12) and (2-11). Changes in the pressure involved may be found from equation (2-14), provided that the equation of state is known or that the temperature can be determined.

In general, under any given set of circumstances four modes of infinitesimal wave propagation are possible at each frequency, although the phase velocity of some modes may be imaginary. The simplest situation is obtained with no external magnetic field. The modes are then of two types, electromagnetic and electrostatic. In the familiar electromagnetic waves, **E** is perpendicular to the direction of propagation. There are two such modes, corresponding to the two directions of polarization. The electrons in a plasma interfere with these transverse waves and increase the wave velocity. If the frequency is less than a certain critical value, which is called the "plasma frequency" and which increases with increasing density, the phase velocity becomes imaginary and electromagnetic waves cannot propagate in the absence of a magnetic field.

The other two modes in the absence of a magnetic field are of electrostatic type, in which **j** and **E** are parallel to the direction of propagation. In one of these modes the positive ions are essentially unaffected and only the electrons oscillate; these oscillations are called "electron waves" or "plasma waves." In the other mode, called "positive-ion waves," the positive ions and electrons generally move together; the inertia of the positive ions determines the wave velocity, which is normally less than for electron waves. In the absence of a magnetic field, the phase velocity of the electron waves becomes imaginary for frequencies less than the plasma frequency, while the positive-ion waves do not propagate above a cut-off frequency equal to $(m_e/m_i)^{1/2}$ times the plasma frequency.

In the presence of a magnetic field these four modes are

profoundly modified, but the number of independent modes remains the same. The term "hydromagnetic wave" is frequently given to the waves which arise in a magnetic field at a frequency small compared to ω_{ci}, the cyclotron frequency of the positive ions. In a hydromagnetic wave the positive ions provide the inertia of the oscillation while the restoring forces are largely magnetic, resulting from the $\mathbf{j} \times \mathbf{B}$ term in equation (2-11). The oscillations may be regarded as waves in the lines of force, which behave as stretched strings, subject to mutual repulsion, and which are "loaded" with the charged particles.

In general, however, a wave in a magnetic field involves both electrostatic and magnetic forces. A high-frequency disturbance, for example, is usually a combination of a transverse electromagnetic wave with a longitudinal electrostatic wave. A hydromagnetic wave travelling across the magnetic field may involve electrical forces similar to those found in electrostatic oscillations. Density gradients may produce a coupling between different types of waves. All these complications are direct results, of course, of the basic equations given in the preceding chapter.

In this chapter we shall investigate the solution of equation (3-1) on the assumption that $\mathbf{E}$ varies as $\exp i(\kappa x - \omega t)$. Since the basic equations have been linearized, the behaviour of any disturbance may be determined from the properties of the sinusoidal oscillations into which it may be decomposed. The angular frequency, ω, equals 2π times the frequency ν, and the "propagation constant" κ equals $2\pi/\lambda$, where λ is the wavelength. Equation (3-1), together with the other equations, then gives a relation between κ and ω. Since the phase velocity V is given by

$$V = \frac{\omega}{\kappa} \tag{3-3}$$

we obtain a functional relation between V and ω, called a "dispersion relation." The group velocity, equal to $d\omega/d\kappa$, may also be found from the dependence of κ on ω.

Knowledge of V is useful in a number of ways. Since the index of refraction varies as $1/V$, the value of V may be used to compute the bending of a ray, when V is a slowly varying function of position. Also, the fraction of energy reflected at an interface depends on the change of V. If the incident ray is travelling at a phase velocity V_1, and strikes at normal incidence a surface beyond which the velocity is V_2, then the fraction R of energy reflected is given by

$$R = \left\{\frac{V_1 - V_2}{V_1 + V_2}\right\}^2 \tag{3-4}$$

provided that the wave amplitude and its first derivative are continuous across the interface, an assumption which is usually valid for normal incidence, and provided also that the energy varies as the square of the amplitude. Finally, the dispersion relation gives the distance over which waves of a particular frequency are damped appreciably. If the distance over which the amplitude decreases by $1/e$ is denoted by d, which we shall call the attenuation distance, then we have

$$\frac{1}{d} = I(\kappa) = I\left(\frac{\omega}{V}\right) \tag{3-5}$$

where $I(\kappa)$ denotes the imaginary part of κ.

3.1 Electromagnetic Waves with No Magnetic Field

When no material current is present, the usual wave equation is obtained from equation (3-1) with $\mathbf{j}$ and σ set equal to zero. In a plasma $\partial\mathbf{j}/\partial t$ must be found from equation (2-12). In this section we shall assume that $\mathbf{B}$ vanishes and shall consider waves in which $\mathbf{j}$ is parallel to the wave front, in which case $\nabla\cdot\mathbf{j}$ vanishes, no charges accumulate, and σ vanishes. In this situation the pressure does not change during the oscillation, and since the undisturbed plasma is assumed uniform we may ignore ∇p_e and ∇p_i throughout. In addition, we shall first set η equal to zero, exploring later the effects introduced by resistivity. Under these conditions **E** is the only nonvanishing term on the right-hand side of equation (2-12). We consider a wave in which E_z is propagated along the x axis, and obtain

$$\frac{\partial^2 E_z}{\partial x^2} - \frac{4\pi\rho Z e^2 E_z}{m_i m_e c^2} = \frac{1}{c^2}\frac{\partial^2 E_z}{\partial t^2} \tag{3-6}$$

With a solution of the form exp $i(\kappa x - \omega t)$ the dispersion relation becomes

$$V^2 = \frac{c^2}{1 - \omega_p^2/\omega^2} \tag{3-7}$$

where the "plasma frequency," ω_p, is defined as

$$\omega_p = \left(\frac{4\pi n_e e^2}{m_e}\right)^{1/2} \times \left(1 + Z\frac{m_e}{m_i}\right)^{1/2} \tag{3-8}$$

The m_e/m_i term may usually be ignored, and numerically we have

$$\nu_p = \frac{\omega_p}{2\pi} = 8.97 \times 10^3 n_e^{1/2} \tag{3-9}$$

As already pointed out, there exist two independent modes of these electromagnetic waves. For a wave travelling in the x direction these modes are either the two plane-polarized modes, with **E** in the y or z directions, respectively, or the two circularly polarized modes.

Evidently in these waves the electric and magnetic fields have the same general nature as in a vacuum, and the electrons oscillate in the electric field; the motion of the positive ions is unimportant. The material current associated with the electron motion modifies the oscillating magnetic field and affects the wave velocity, V. In fact, for ω less than ω_p, V and κ become imaginary, and the wave does not propagate. The reason for this result may be seen directly from the basic equations. Equation (2-12) shows that $\partial \mathbf{j}/\partial t$ and **E** must have the same sign and phase. Since both **j** and **E** vary sinusoidally in time, **j** and $\partial \mathbf{E}/\partial t$ will be 180° out of phase, which means that they will be proportional to each other but of opposite signs. As a result the material current opposes the displacement current in equation (2-19). At frequencies well above the plasma frequency the electron inertia keeps the material current low and the wave is relatively unaffected. At lower frequencies the material current will more than cancel the displacement current, reversing the sign of the total current. As may be seen from equation (2-19) this has qualitatively the same effect as changing c to an imaginary number.

When V is imaginary, the wave will penetrate some distance into the plasma, but the amplitude will decrease by a factor $1/e$ in a distance d. From equations (3-5) and (3-7) we find

$$d = \frac{c}{\omega_p} \times \frac{1}{(1 - \omega^2/\omega_p{}^2)^{1/2}} \tag{3-10}$$

Thus for ω much less than ω_p, the value of d approaches $1/2\pi$ times the wavelength, *in vacuo*, for radiation at the plasma frequency.

When the finite resistivity is taken into account, κ^2 becomes complex. For ω sufficiently large compared to ω_p the wave propagates as before, but is gradually damped out, the energy being dissipated in the ohmic losses ηj^2. If ω is sufficiently small compared to ω_p, the term in $\partial \mathbf{j}/\partial t$ may be neglected in equation (2-12), and hence $\mathbf{j}$ may be set equal to $\mathbf{E}/\eta$ in equation (3-1). We then obtain

$$\kappa^2 = \frac{\omega^2}{c^2}\left(1 + \frac{4\pi i c^2}{\omega\eta}\right) \tag{3-11}$$

When $4\pi c^2/\omega\eta$ is large compared to unity, κ^2 is predominantly imaginary. From equations (3-5) and (3-11) we find that the attenuation distance, d, is then given by the usual formula for the skin effect

$$d = \left(\frac{\eta}{2\pi\omega}\right)^{1/2} \tag{3-12}$$

For an electron-proton gas the resistivity is given by equation (5-37), provided that the angular frequency, ω, is much less than $1/t_c$, where t_c is the self-collision time; for this condition equation (3-12) yields

$$d = \frac{4.07 \times 10^5}{T^{3/4}} \left(\frac{\ln \Lambda}{\nu}\right)^{1/2} \tag{3-13}$$

where ν is again the frequency. The condition that $4\pi c^2/\omega\eta$ be large implies that equation (3-13) is valid only if it gives a value of d much less than c/ω, $1/(2\pi)$ times the wavelength in free space. Also, the neglect of the $\partial \mathbf{j}/\partial t$ term in equation (2-12) is valid only if equation (3-13) yields a greater value than equation (3-10); in general, of the two equations (3-10) and (3-13), the greater value for d should be taken.

3.2 Electrostatic Waves with No Magnetic Field

We investigate now waves in which the restoring force is electrostatic, the electric charge resulting from the divergence of the current density $\mathbf{j}$. As already noted the two basic modes are electron waves and positive-ion waves.

In analyzing the electron waves, we may ignore the ∇p_i term in equation (2-12). If we set η as well as $\mathbf{B}$ equal to zero, the generalized Ohm's law then yields

$$\frac{m_i m_e c^2}{Z\rho e^2} \frac{\partial \mathbf{j}}{\partial t} = \mathbf{E} + \frac{cm_i}{eZ\rho} \nabla p_e \tag{3-14}$$

The term ∇p_e in equation (3-14) is based on the assumption that the stresses associated with the wave are isotropic. This assumption is not valid here, since the frequencies of oscillation involved are large compared to the collision frequencies, and the oscillating part of the stress tensor will be anisotropic. Nor can we replace p_e by $p_{e\perp}$, since no magnetic field is present. However, Oberman has shown (quoted in ref. (9)) that equation (3-14) can still be used if we take for p_e the component of the stress tensor in the direction of wave propagation, provided that V, the phase velocity of the wave, is large compared to the random velocities of any particles present. Their analysis indicates that with such a large phase velocity the heat flow in any direction is negligible, and the component of the stress tensor in any direction will change adiabatically. Since the compression in the present case is one-dimensional, we may take γ equal to 3, in accordance with the discussion in Section 1.4. If we neglect terms of order m_e/m_i, comparable to the ∇p_i term already ignored, we obtain

$$\nabla p_e = 3kT\nabla n_e = -\frac{3kT_e c}{e}\nabla\sigma; \tag{3-15}$$

If now we take the divergence of equation (3-14) and express $\mathbf{j}$, $\mathbf{E}$, and ∇p_e in terms of σ by means of equations (2-15), (2-16), and (3-15) we obtain

$$\frac{\partial^2\sigma}{\partial t^2} = -\omega_p{}^2\sigma + \frac{3kT_e}{m_e}\nabla^2\sigma \tag{3-16}$$

where ω_p is again given in equation (3-8). This procedure eliminates modes in which $\mathbf{j}$ shows no divergence; equation (3-2) is satisfied automatically, since this relation follows from equations (2-15) and (2-16). If we assume that σ varies as exp $i(\kappa x - \omega t)$, the velocity becomes

$$V^2 = \frac{\omega^2}{\kappa^2} = \frac{\omega_p{}^2}{\kappa^2} + \frac{3kT_e}{m_e} \tag{3-17}$$

In the limit of very low temperature, these electron oscillations can have only one frequency, ω_p, regardless of wavelength, a result derived originally by Tonks and Langmuir (39). In this limiting case the electrons oscillate under the restoring force resulting from charge separation; the oscillation frequency may be derived approximately from the equations of motion and of continuity for the electrons and from Poisson's Law, equation (2-16).

To express V as a function of ω for these electron waves we may eliminate κ^2 from equation (3-17) and obtain

$$V^2 = \frac{1}{1 - \omega_p{}^2/\omega^2}\frac{3kT_e}{m_e} \tag{3-18}$$

Evidently the dispersion relation for electron waves, when **B** vanishes, is similar to equation (3-7) for electromagnetic waves, except that the root mean square electron thermal velocity replaces c. Since we have already assumed that V^2 much exceeds $3kT_e/m_e$, equation (3-18) is valid only for ω relatively close to ω_p; this restriction is equivalent to requiring that the wavelength be much greater than the Debye shielding distance.

Electron oscillations may also be analyzed in terms of the actual distribution of velocities, as in the basic works by Landau (27) and by Bohm and Gross (10). These analyses, which are based on the Boltzmann equation, give dispersion relations which are usually not much different from those obtained by use of the macroscopic equations; in addition, these more detailed theories give results on wave excitation and damping, not obtainable from the macroscopic equations. Some of these results are summarized briefly in Section 3.5.

Observations of electron oscillations were first obtained by Penning (31); since then a number of workers have detected these oscillations. In particular, detailed observations by Looney and Brown (28) show that the frequency of these oscillations is about equal to the plasma frequency ν_p given in equation (3-9). A survey of experimental and theoretical research in this field has been given by Gabor (18).

The enhanced radio noise emitted from active regions on the sun is probably produced in part by electron oscillations. The coupling between these oscillations inside a plasma and the electromagnetic radiation outside has been analyzed by Field (16), Tidman (38), and others. These analyses indicate that an appreciable amount of oscillation energy can be radiated into space.

We proceed now to a consideration of the positive-ion oscillations. These are waves of relatively low frequency in which electrical neutrality is preserved to a high degree. Because of the low frequency we may assume initially that ω is less than the collision frequencies for electrons and positive ions; the stress tensor is then isotropic because of short mean free path. Since the positive ions are moving, $\mathbf{v}$ oscillates and we must use the equation of motion, (2-11) instead of equation (2-12). The change of p, which we denote by $p^{(1)}$, may be determined from the adiabatic equation of state, with the electron and ion temperatures assumed equal; $\rho^{(1)}$, the change of ρ, is determined from the equation of continuity, (2-14). If we combine these two equations, and let ρ equal $n_i m_i$, we obtain

$$\frac{1}{\rho}\frac{\partial p^{(1)}}{\partial t} = -\frac{(1 + Z)\gamma kT}{m_i}\nabla \cdot \mathbf{v} \tag{3-19}$$

We now differentiate equation (2-11) with respect to time, and take the gradient of equation (3-19), eliminating $\nabla \partial p^{(1)}/\rho \partial t$; we find, for a wave travelling in the x direction,

$$\frac{\partial^2 v_x}{\partial t^2} = \frac{(1 + Z)\gamma kT}{m_i}\frac{\partial^2 v_x}{\partial x^2} \tag{3-20}$$

Products of the first order quantities, $\mathbf{v}$, $p^{(1)}$, or $\rho^{(1)}$, have been ignored in this analysis. Equation (3-20) is the usual result for an acoustic wave. To replace γkT by the more detailed average over electrons and positive ions, we must determine n_i and n_e from separate equations of continuity for each type of particle, and the pressures from separate adiabatic relations. If we assume in these computations that electrical neutrality is nearly preserved in the perturbation and that in consequence $n_e^{(1)}$ about equals $Zn_i^{(1)}$, we find that the velocity of this acoustic wave becomes

$$V^2 = V_S^2 \equiv \frac{Z\gamma_e kT_e + \gamma_i kT_i}{m_i} \tag{3-21}$$

Since this velocity of positive-ion waves occurs in various other analyses, we use the symbol V_S for this quantity. Evidently V_S is the conventional sound speed; positive-ion waves may be regarded as a form of acoustic wave.

The electric field E_x, while it does not appear in the above equations, actually plays an important part in these oscillations. In the macroscopic equation (2-11) the gradient of the total pressure appears, and both T_i and T_e appear in equation (3-21) for V_S. The importance of T_i and of the positive-ion pressure is obvious, but it is perhaps not so clear how the electrons affect the wave, especially if electron-ion collisions are not considered. The electron motion is essentially determined by the generalized Ohm's Law, equation (2-12). If we substitute from equation (3-2) we find that the ratio of the $\partial \mathbf{j}/\partial t$ term in equation (2-12) to $\mathbf{E}$ is simply $(\omega/\omega_p)^2$. Hence for low-frequency oscillations the $\partial j/\partial t$ term, which represents primarily the effect of electron inertia, can be set equal to zero, and the two dominant terms on the right-hand side, $\mathbf{E}$ and the ∇p_e term, must be nearly equal and opposite. The electric field then transmits to the positive ions the force associated with the electron pressure gradient. As a result if T_e much exceeds T_i, the velocity of these acoustic waves is much greater than the thermal velocity of the positive ions, and is about equal to the random velocity the positive ions would have if T_i were as great as T_e.

To explain the situation in another way, we may consider that the positive ions move so slowly that the electrons at each time have their equilibrium distribution in the electrostatic potential field U. Thus we may write

$$n_e = \bar{n}_e e^{eU/ckT} \tag{3-22}$$

where $\bar{n}_e$ is a mean density. Since we have assumed that electron-electron collisions are very rapid in comparison with the oscillations under consideration, equation (3-22) is clearly valid. As a result of this equation, the ∇p_e term in equation (2-12) equals $-\mathbf{E}$. For any distribution of positive ions the potential U will adjust itself so that n_e, given by equation (3-22), is nearly equal to Zn_i. The gradient of this potential then provides an electrical field which gives a restoring force on the positive ions. If it were not for the electrical forces, n_e and n_i would still vary together because of electron-ion collisions, as in an acoustic wave in an ordinary gas. However, the electrical forces in a positive ion wave keep the value of n_e much more closely equal to Zn_i than collisions alone could do.

Since the acoustic velocity V_S is less than the electron

thermal velocity, one cannot assume, as with electron oscillations, that the phase velocity much exceeds the random velocities of any particles present. Hence these results are no longer valid when the collision frequencies are less than ω. While an exact solution for progressive acoustic waves has been obtained by Bernstein, Greene, and Kruskal (8), the full variety of phenomena that become possible in this limit of weak collisions has not been fully explored, though it seems likely that equation (3-21) will still be approximately correct for the real part of the wave velocity. The theoretical uncertainty is greatest if T_i is about equal to T_e, since then the random velocities of the positive ions are about equal to V. As shown by Fried and Gould (17), the linearized Boltzmann theory predicts strong damping in this case, when collisions are weak; this "Landau damping" is discussed in Section 3.5. If T_i is much less than T_e, equation (3-21) should be valid provided that the electron velocity distribution is kept Maxwellian by collisions or by some other process.

Let us consider in more detail the dispersion relation if we relax the assumption of electrical neutrality in the macroscopic equation of motion, and compute $n_e^{(1)}$ and $n_i^{(1)}$ separately. We assume here that T_i/T_e is negligible so that damping can be ignored. Again we shall ignore $\mathbf{B}$ and η, in addition to m_e/m_i and p_i. The results will then be applicable to those situations where the wavelength is less than the Debye shielding distance. Equations (2-11) and (2-12) are still valid under these conditions; the correction terms in $n_e - Zn_i$ are all negligible in the present situation. Alternatively, one may take separate macroscopic equations for electrons and ions separately, and apply equation (3-2) by means of equation (2-9). In either case we obtain a dispersion relation in which two roots appear. One describes the electron oscillations, and yields equation (3-17). The other, in which we are interested here, gives for the velocity of the positive-ion wave

$$V^2 = \frac{Z\gamma_e kT_e}{m_i} \frac{1}{1 + \gamma_e \kappa^2 h^2} \tag{3-23}$$

where h, the Debye shielding distance, is given in equation (2-3). For κh small equation (3-23) leads to the previous result, equation (3-21). If κh is large, equation (3-23) is most conveniently put in the form

$$\omega^2 = \frac{Zm_e\omega_p^2}{m_i}\frac{1}{1 + 1/(\gamma_e\kappa^2h^2)} \tag{3-24}$$

Evidently for T_e sufficiently large κh much exceeds unity, and the frequency approaches a constant value, obtained by substituting m_i/Z for m_e in equation (3-8) for ω_p. In this limit the stress tensor for the electrons disappears from the problem, and equation (3-24) is valid even in the absence of collisions, provided that the number of electrons moving at the phase velocity of the wave is negligible. In this situation the electrons provide a medium of constant density which neutralizes the electrical charges of the positive ions in the undisturbed plasma; the ion oscillations are then unshielded and the frequency rises to essentially the corresponding plasma frequency for positive ions.

Oscillations at the frequency $\omega_p(m_e/m_i)^{1/2}$ have been observed by Hernqvist (23) in a plasma consisting of electrons at high energy, with singly charged positive ions at essentially room temperature. The acoustic waves described above, with a velocity nearly independent of frequency, have recently been observed by Alexeff and Neidigh (1).

3.3 Hydromagnetic Waves

When a magnetic field is present in a plasma, the four modes analyzed in the preceding two sections are modified. The analysis in the general case is straightforward but detailed. To bring out clearly the properties of some of the more important types of waves we give here a simplified analysis of hydromagnetic waves, defined as the disturbances which propagate through a plasma when ω is much less than ω_{ci}. As in the preceding sections, we shall assume that the undisturbed plasma is uniform, with the gravitational potential ϕ equal to zero.

A simple type of hydromagnetic wave, which was first analyzed by Alfvén (2) and which we shall call an Alfvén wave, is found when the velocity **v** and the change $\mathbf{B}^{(1)}$ in the magnetic field are both parallel to the y axis, but independent of y, while the mean magnetic field **B** is parallel to the x axis, which is also the direction of propagation. The current density **j** is then parallel to the z axis; both **j** and **v** are parallel to the wave front. In this type of motion the acceleration in the y direction is unaffected by the pressure, since there can be no gradient parallel to the wave front. We neglect the following terms in equation (2-12): (a) $\partial \mathbf{j}/\partial t$; (b) $\mathbf{j} \times \mathbf{B}$; (c) $\eta \mathbf{j}$. The first two of these terms are of order $\omega^2/\omega_{ci}\omega_{ce}$ and ω/ω_{ci}, respectively, with respect to the terms retained, and are unimportant if ω is sufficiently small. With these simplifications, equations (2-11) and (2-12) become

$$\rho \frac{\partial v_y}{\partial t} = j_z B \tag{3-25}$$

$$E_z - v_y B = 0 \tag{3-26}$$

If we express j_z in terms of E_z, by means of equation (3-25) and (3-26), and substitute into equation (3-1), we obtain

$$\frac{\partial^2 E_z}{\partial x^2} = \left(1 + \frac{4\pi\rho c^2}{B^2}\right)\frac{1}{c^2}\frac{\partial^2 E_z}{\partial t^2} \tag{3-27}$$

Equation (3-27) is the wave equation for a medium of dielectric constant K, where K is again given by equation (2-32). Thus an Alfvén wave may be regarded as a normal electromagnetic wave, modified by the high dielectric constant of the gas. The phase velocity is given by

$$V = \frac{c}{K^{1/2}} = \frac{c}{(1 + 4\pi\rho c^2/B^2)^{1/2}} \tag{3-28}$$

For K large compared to unity, this velocity is about equal to the Alfvén speed, V_A, defined by the relation

$$V_A = \frac{B}{(4\pi\rho)^{1/2}} \tag{3-29}$$

From another standpoint it is convenient to regard an Alfvén wave as the vibration of a line of magnetic force; according to the discussion in Section 2.5 the plasma will move with the lines of force. The stress system associated with the magnetic field is a tensile stress, $B^2/8\pi$, along the lines of force and a compressive stress, $B^2/8\pi$, between the lines of force. As pointed out by Cowling (13), this system is equivalent to a hydrostatic pressure equal to $B^2/8\pi$ and a doubled tensile stress $B^2/4\pi$. If $y(x)$ is the displacement of the medium, the restoring force per cubic centimeter equals the tensile stress times $\partial^2 y/\partial x^2$, as shown in the theory of vibrating strings. Hence the equation of motion becomes

$$\rho \frac{\partial^2 y}{\partial t^2} = \frac{B^2}{4\pi} \frac{\partial^2 y}{\partial x^2} \tag{3-30}$$

which agrees with equation (3-27) for K large. Since an Alfvén wave is essentially a transverse electromagnetic wave, it can be polarized in either of two ways, giving two independent modes of propagation parallel to **B**. The other two modes are the electron and positive-ion waves, which have previously been analyzed and which are unaffected by **B** for this direction of propagation.

For large-amplitude plane-polarized Alfvén waves a ponderomotive force $-j_z B_y^{(1)}$ in the x direction, proportional to the square of the wave amplitude, must be taken into account. In an incompressible fluid no velocities can appear in the x direction, since $\partial v_x/\partial x$ must vanish by equation (2-14); in this incompressible case the force in the x direction will be balanced by a corresponding pressure gradient, and equations (3-27) and (3-28) are valid for waves of arbitrary amplitude. In a compressible gas velocities in the x direction will appear for large amplitude, modifying equations (3-27) and (3-28). Moreover, if the collision rate is low, and the particle velocities are comparable with the wave velocity, V, other complications appear for a finite-amplitude plane wave. However, for a circularly polarized wave the situation becomes much simpler. As shown by Ferraro (15), $\mathbf{j} \times \mathbf{B}^{(1)}$ disappears for circularly polarized Alfvén waves, and equation (3-28) gives the correct velocity for such waves of large amplitude. Moreover, the distribution of particle velocities now has no effect, provided only that the pressure is isotropic—see Section 4.2.

Alfvén waves may be generated by an initial displacement of material perpendicular to **B**. In general, such a displacement will produce waves going out in each direction along the lines of force. Further detailed properties of these waves, including the damping produced by the finite resistivity, are discussed in Alfvén's book (3).

We next consider a wave in which the particle velocity, $\mathbf{v}$, is parallel to the direction of propagation, both being perpendicular to **B**. This is a longitudinal hydromagnetic wave, which we shall call a "magnetosonic" wave. In this case the pressure gradient must be retained in equation (2-11), and equation (3-25) must now be modified accordingly. Since conditions are uniform along the lines of magnetic force, the conditions are satisfied for replacing $\nabla \cdot \boldsymbol{\Psi}$ by $\nabla p_\perp$ in the macroscopic equations. Hence the analysis of magnetosonic waves is valid even in the absence of collisions, without any restrictions on the relation between particle velocities and the wave velocity. As before we shall neglect terms of order ω/ω_{ci}; i.e., $\partial \mathbf{j}/\partial t$ and $\mathbf{j} \times \mathbf{B}$, as well as $\eta \mathbf{j}$, are all ignored in equation (2-12). Also, it may be shown that the charge density σ in equation (3-1) is negligible for small ω/ω_{ci}. We obtain, after some algebra

$$V^2 = \frac{V_A^2 + V_S^2}{1 + V_A^2/c^2} \tag{3-31}$$

where V_A and V_S are the Alfvén speed and the acoustic wave, or sound speed, given in equations (3-29) and (3-21), respectively. In the absence of collisions the changing magnetic field affects the particle velocities in the two directions perpendicular to **B**, and the compression is two-dimensional; in this case γ_e and γ_i may be set equal to 2 in equation (3-21).

In these disturbances the inertia of the positive ions is

opposed by two restoring forces—the pressure gradient of the gas and the gradient of the compressional stresses between the lines of force. If the magnetic "pressure," $B^2/8\pi$, is large compared to the material pressure, p, the velocity of the magnetosonic wave is about the same as the Alfvén speed, although Alfvén waves and magnetosonic waves involve quite different types of magnetic stresses. If the material pressure is much greater, the compressional wave is essentially an acoustic wave, similar to the positive-ion wave discussed in the previous section.

When a hydromagnetic wave is moving at an angle, θ, with respect to the lines of force three modes of oscillation are possible; the restriction to low frequencies eliminates the electron waves, which normally do not appear for frequencies less than ω_p. The analyses by Herlofson (22) and van de Hulst (40) indicate that one of these three modes is an Alfvén wave, with a velocity perpendicular both to the magnetic field, **B**, and the propagation vector, $\boldsymbol{\kappa}$. If one considers the magnetic field as a group of stretched strings, it is evident that any disturbance in which the displacement is perpendicular both to **B** and to the wave front will move parallel to **B** at the usual Alfvén speed. Thus the apparent velocity of the wave, measured normal to the wave front, becomes

$$V = \frac{c \cos \theta}{K^{1/2}} \approx \frac{B}{(4\pi\rho)^{1/2}} \cos \theta \qquad (3\text{-}32)$$

where B is the magnetic field in the undisturbed plasma and θ is the angle between the magnetic field and the direction of propagation. Mathematically equation (3-32) is readily derived by considering that the mean magnetic field has a component, B_z, perpendicular to the direction of propagation, in addition to the parallel component, B_x, appearing in equations (3-25) through (3-28). Additional components j_x and E_x parallel to the propagation direction must also be introduced, related through equation (3-2). While the longitudinal electrical field is a new feature, $\mathbf{v}$ remains parallel to the y axis, the motion remains incompressible for small amplitude, and equation (3-28) is replaced by equation (3-32), with the square of the total magnetic field, $B_x^2 + B_z^2$, appearing in K. It is readily verified that the Poynting vector is parallel to **B**; as would be expected physically for this type of disturbance, the energy flows along the lines of force with the velocity $c/K^{1/2}$.

The other two modes involve macroscopic motions in the plane defined by **B** and $\boldsymbol{\kappa}$. Both these modes involve some compression of the gas. For K large compared to one, yielding an Alfvén speed small compared to c, we have

$$\frac{1}{V^2} = \frac{(V_A^2 + V_S^2)}{2V_A^2V_S^2}\sec^2\theta \times \left\{1 \pm \left(1 - \frac{4V_A^2V_S^2\cos^2\theta}{(V_A^2 + V_S^2)^2}\right)^{1/2}\right\} \tag{3-33}$$

where V_A and V_S are again given in equations (3-29) and (3-21). When θ is near 90° the square root may be expanded, and if the minus sign is taken equation (3-33) yields V^2 equal to $V_A^2 + V_S^2$, the magnetosonic velocity found in equation (3-31). For θ equal to zero this same mode becomes either an acoustic wave ($V = V_S$) or an Alfvén wave ($V = V_A$), whichever has the greater velocity. The other mode, corresponding to a plus sign, yields a velocity which vanishes as $\cos\theta$, as with the Alfvén wave. We shall call a "modified Alfvén wave" that mode which is a pure Alfvén wave at zero θ only. The variation of V with θ for two different values of V_S/V_A is shown in Figure 3.1. It is evident from this figure that the mode with the highest velocity (called the "fast wave" by van de Hulst) is a modified Alfvén wave if V_S is less than V_A. Similarly, the wave of lowest velocity is called the "slow wave." For V_S/V_A very large the modified Alfvén wave, which is now the slow wave, is scarcely distinguishable from the pure Alfvén wave. In this same limit the fast wave is an acoustic wave, with a velocity nearly independent of θ.

The damping of all three hydromagnetic modes, as a result of viscosity and finite resistivity, has been evaluated by van de

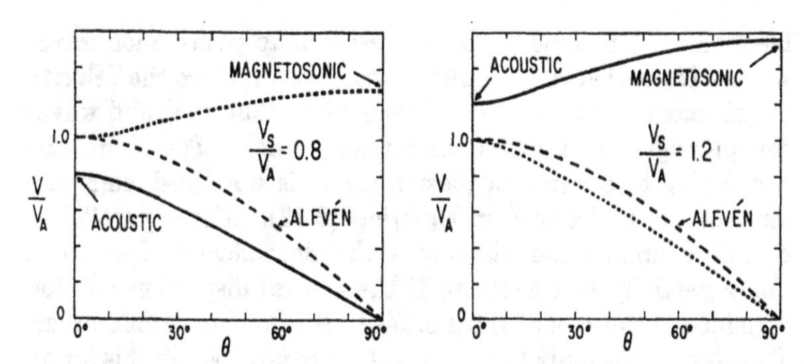

Figure 3.1. Hydromagnetic wave velocity. The ratio of the phase velocity, V, to the Alfvén speed, V_A, is plotted as a function of θ, the angle between the propagation vector and the magnetic field, for two ratios of sound speed V_S to V_A. The dashed line represents an Alfvén wave, while the dotted line depicts a modified Alfvén wave.

Hulst (40) and others (9). When ω approaches ω_{ci}, the polarization of the normal modes is changed from plane to elliptical; the dispersion relation under these conditions is discussed in the following section. A general survey of hydromagnetic phenomena, including steady motions and other effects as well as low-frequency waves, has been given by Lundquist (29).

3.4 Waves in a Cold Uniform Plasma

We consider now the general theory of wave propagation in a plasma with a magnetic field present. As before, the wave will be considered infinitesimal, and all quantities in the unperturbed plasma will be taken to be uniform. While the general theory for warm plasmas is straightforward, the analysis is cumbersome. The chief effects of temperature occur for the positive-ion and hydromagnetic waves already discussed. Since in any case the effects produced by random velocities are not always correctly given by the macroscopic equations, we shall simplify the analysis by neglecting both p_i and p_e in the macroscopic equations. This procedure is valid for a cold plasma.

This approximation naturally reduces the number of possible modes. The mode which corresponds to positive-ion waves when $\boldsymbol{\kappa}$ is parallel to **B** is eliminated entirely, since the velocity of this mode varies as $T^{1/2}$. It is evident that no sound waves can propagate in a gas at zero temperature. The mode corresponding to electron or plasma waves is simplified, since the term in ∇p_e is dropped in equation (2-12). For $\boldsymbol{\kappa}$ parallel to **B** this simplification eliminates the dependence of ω on κ. For $\boldsymbol{\kappa}$ partially transverse to **B** the general dispersion relation is reduced from one of third order in κ^2 to one in second order. Thus for each ω^2 only two values of κ^2 are possible; in this sense, only two modes are present. Since ω^2 still occurs in the third order in the general dispersion relation, however, there are still three values of ω^2 possible for each value of κ^2, and in this sense the longitudinal electron waves are retained. As we shall see below, these electron waves appear in a particular range of ω for transverse propagation.

For a plane wave travelling at an arbitrary angle, θ, with respect to the magnetic field the formulae become somewhat complicated. This general case has been discussed by Aström (5). More recently a very complete treatment has been given by Allis, Buchsbaum, and Bers (4), much of whose analysis we shall follow here. Most of the discussion will be limited to the two special cases θ equals $\pi/2$, ($\boldsymbol{\kappa}$ perpendicular to **B**) and θ equals 0 ($\boldsymbol{\kappa}$ parallel to **B**), with some results for general θ given at the end of this section.

a. Propagation across **B**. We assume that the unperturbed magnetic field, **B**, is in the z direction, while the wave propagates in the x direction. Equations (2-11), (2-12), (3-1), and (3-2) may be combined; if we consider that **j**, **E**, and **v** all vary as $\exp i(\kappa x - \omega t)$, we find that j_z and E_z are not coupled with the other components. From equations (2-12) and (3-1) we can determine the phase velocity for a wave involving only j_z and E_z, in which case we recover equation (3-6). This mode, which is called the ordinary mode, is a transverse wave whose electric vector is parallel to **B** in the undisturbed plasma, and which is, therefore, entirely independent of **B**.

The components of **E** and **j** in the other two directions are coupled together in a single wave, which we call the extraordinary mode. Since both E_x and E_y differ from zero, the wave is partly transverse, partly longitudinal. With straightforward algebra we obtain two simultaneous equations for j_x and j_y. The condition that these equations possess a solution is that the determinant of the coefficients must vanish. This condition yields the dispersion relation

$$\frac{c^2}{V^2} = 1 - \frac{\omega_p^2}{\omega^2 - \omega_{ce}\omega_{ci} + \dfrac{\omega^2(\omega_{ce} - \omega_{ci})^2}{\omega_p^2 - \omega^2 + \omega_{ce}\omega_{ci}}} \tag{3-34}$$

As B approaches zero, the cyclotron frequencies ω_{ce} and ω_{ci} [defined in equation (1-2)] approach zero, and we again recover equation (3-7). As ω approaches zero we obtain the usual result, equation (3-28), for hydromagnetic waves, with the aid of the identity

$$\frac{\omega_{ce}\omega_{ci}}{\omega_p^2} = \frac{V_A^2}{c^2} = \frac{1}{K - 1} \tag{3-35}$$

where the Alfvén speed, V_A, is defined in equation (3-29) and the dielectric constant, K, in equation (2-32). In this limiting case the extraordinary wave is a magnetosonic wave rather than an Alfvén wave, but since the temperature has been ignored the velocity equals the Alfvén speed. More generally, for ω/ω_p small and for V_A/c also small, equation (3-34) yields, if we neglect ω_{ci}/ω_{ce} as compared to unity,

$$\frac{V^2}{V_A^2} = 1 - \frac{\omega^2}{\omega_{ce}\omega_{ci}}\left(1 + \frac{\omega_{ce}^2}{\omega_p^2}\right) \tag{3-36}$$

Finally, if ω is comparable to or greater than ω_{ce}, we may ignore ω_{ci} in equation (3-34) and obtain the usual dispersion relation for electromagnetic waves in the ionosphere—see the discussion by Mitra (30) and Ratcliffe (33).

It is frequently helpful, in analyzing a dispersion relation, to examine the frequencies at which V is zero or infinity. The former are called "resonances," since these are the frequencies at which a plasma will be in resonance with an applied oscillating transverse electric field. The latter are called "cut-offs." At a cut-off a wave is usually reflected, while at a resonance either absorption or reflection may occur depending (36) on the nature of the damping processes involved. It is evident from equation (3-7) that the ordinary wave has no resonance, but possesses a cut-off at ω equal to ω_p. This cut-off, which characterizes the ordinary mode for any direction of propagation (except for the singular direction θ equal to 0), is called the "plasma cut-off."

The extraordinary mode has two resonances and two cut-offs. The resonant frequencies may be found from equation (3-34) on setting $V = 0$ and solving for ω. If we expand the solution in powers of m_e/m_i, and retain only the lowest significant terms, we obtain

$$\omega^2 = \begin{cases} \omega_{h1}^2 \left(\dfrac{\omega_p^2 + \omega_{ce}\omega_{ci}}{\omega_p^2 + \omega_{ce}^2} \right) \\ \omega_{h2}^2 \end{cases} \tag{3-37}$$

where ω_{h1} and ω_{h2} are the lower and upper hybrid frequencies, defined by

$$\omega_{h1}^2 = \omega_{ce}\omega_{ci} \tag{3-38}$$

$$\omega_{h2}^2 = \omega_p^2 + \omega_{ce}^2 \tag{3-39}$$

The lower resonant frequency approaches the lower hybrid frequency as ω_p^2/ω_{ce}^2 [equal to $(K - 1)m_e/m_i$ from equation (3-35)] becomes large. With decreasing ω_p/ω_{ce}, however, this resonant frequency decreases to $m_e\omega_p^2/m_i$, the ion plasma frequency, and finally decreases to ω_{ci} when K approaches unity.

These resonances occur at the frequencies of free oscillations of a plasma with no electromagnetic field; i.e., with zero electric field parallel to the wave front. Since $\nabla \times \mathbf{E}$ is known from equation (2-18), this condition applies for the short wavelength and large κ characteristic of resonance. At the upper hybrid frequency only the electrons oscillate, and the resonant frequency follows directly from equation (2-12), with $\mathbf{v}$, p_e, p_i, and η neglected and equation (3-2) used to determine E_x, the electrostatic field in the direction of wave propagation; E_y is set equal to zero, although as a result of the j_xB_z term a current component j_y must be taken into account. The resonant oscillations represent the joint influence of electrostatic and magnetic forces on electrons gyrating in the xy plane, perpendicular both to $\mathbf{B}$ and the wave front.

At the lower hybrid frequency, discussed by Auer, Hurwitz, and Miller (6), electrons and ions oscillate together; electrical neutrality requires that their velocities in the direction of propagation, taken to be along the x axis, be about equal. If the magnetic field is in the z direction, both electrons and ions will be subject to a force in the y direction, equal to $-qv_xB_z$. The positive ions, because of their large inertia, will not be affected by this force, but the electrons will be strongly accelerated, and a current j_y will result, proportional to the plasma displacement in the x direction. The ponderomotive force j_yB_z on this current then provides a restoring force on the positive ions. An electrostatic field E_x will appear to keep the electron and ion velocities nearly equal in the x direction, and this field will nearly cancel the force j_yB_z on the electrons and will transmit this force to the ions. Evidently the ion inertia controls the acceleration in the x direction, while the electron inertia controls the current in the y direction, which is responsible for the force in the x direction. Hence, the resonant frequency involves both m_i and m_e.

The two cut-off frequencies may also be obtained from equation (3-34), with V now set equal to infinity. As we shall see below, these cut-offs may be designated as left or right, and designated by ω_{ol} and ω_{or}. We obtain the simple relationships

$$\frac{\omega_p^2}{\omega_{ol}^2} = \left(1 + \frac{\omega_{ce}}{\omega_{ol}}\right)\left(1 - \frac{\omega_{ci}}{\omega_{ol}}\right) \tag{3-40}$$

$$\frac{\omega_p^2}{\omega_{or}^2} = \left(1 - \frac{\omega_{ce}}{\omega_{or}}\right)\left(1 + \frac{\omega_{ci}}{\omega_{or}}\right) \tag{3-41}$$

These equations may be regarded as giving the density at cut-off for any arbitrary ω_{ol} or ω_{or} and for given values of the cyclotron frequencies. If we solve for the frequencies directly, again keeping only the lowest significant terms in m_e/m_i, we obtain

$$\omega_{ol} = \left(\omega_p^2 + \frac{(\omega_{ce} + \omega_{ci})^2}{4}\right)^{1/2} + \frac{\omega_{ci} - \omega_{ce}}{2} \qquad (3\text{-}42)$$

$$\omega_{or} = \left(\omega_p^2 + \frac{\omega_{ce}^2}{4}\right)^{1/2} + \frac{\omega_{ce}}{2} \qquad (3\text{-}43)$$

If $\omega_p^2/\omega_{ce}\omega_{ci}$ is small, ω_{ol} is about equal to ω_{ci}, while ω_{or} is equal to ω_{ce}. Hence these two cut-offs are called the cyclotron cut-offs. For ω_p/ω_{ce} large, however, both ω_{ol} and ω_{or} are nearly equal to ω_p.

These cut-offs and resonances are conveniently portrayed in an Allis diagram (4), where ω_{ce}^2/ω^2 is plotted against ω_p^2/ω^2. For constant ω the vertical and horizontal scales are proportional to B^2 and to n_e, respectively. Figure 3.2 gives such a diagram for low values of the ordinate, relevant for high frequencies. The cut-offs are shown as solid lines, the resonances as dashed lines. Along the dotted line the Alfvén speed, V_A, equals $c(m_e/m_i)^{1/2}$; evidently ω_{ce}/ω_p equals unity along this line. The lower right-hand regions of the figure correspond to ω_p greater than ω_{ce}; i.e., to high density or low magnetic field. The polar diagrams in each region are explained in subsection (c) below. Plots similar to Figure 3.2, without the polar diagrams, have been given by Clemmow and Mullaly (12). Figure 3.3 gives a similar diagram for low frequencies; logarithmic scales have been used to permit giving full information. A mass ratio of 1836 for m_i/m_e has been assumed in Figure 3.3.

The values of V^2/c^2 for the ordinary and extraordinary modes are shown in the lowest diagrams in Figures 3.4 and 3.5, the former for high frequency, (low ω_p^2/ω^2 and low ω_{ce}^2/ω^2), the latter for low frequency. The independent variable in all of these diagrams is taken to be ω_p^2/ω^2; This provides consistency

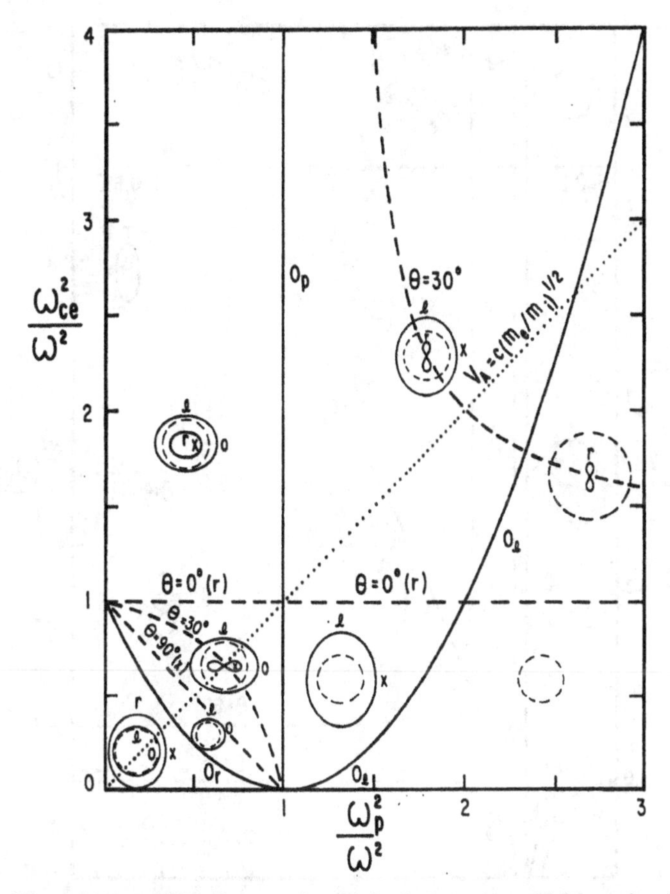

Figure 3.2. Allis diagram for high frequencies. Values of $\omega_p{}^2/\omega^2$ and $\omega_{ce}{}^2/\omega^2$ at which V is infinite (cut-offs) are shown by solid lines; the values at which V vanishes (resonances) for particular values of θ are shown by dashed lines. The small diagrams depict $V(\theta)$ for the various modes of propagation.

with the Allis diagram and shows how V^2/c^2 changes with changing plasma density. The same logarithmic scales are used in Figure 3.5 as in Figure 3.3. To simplify the diagrams in Figure 3.5 the ratio of V^2 to c^2/K (the square of the velocity for Alfvén waves) has been plotted instead of V^2/c^2. In re-

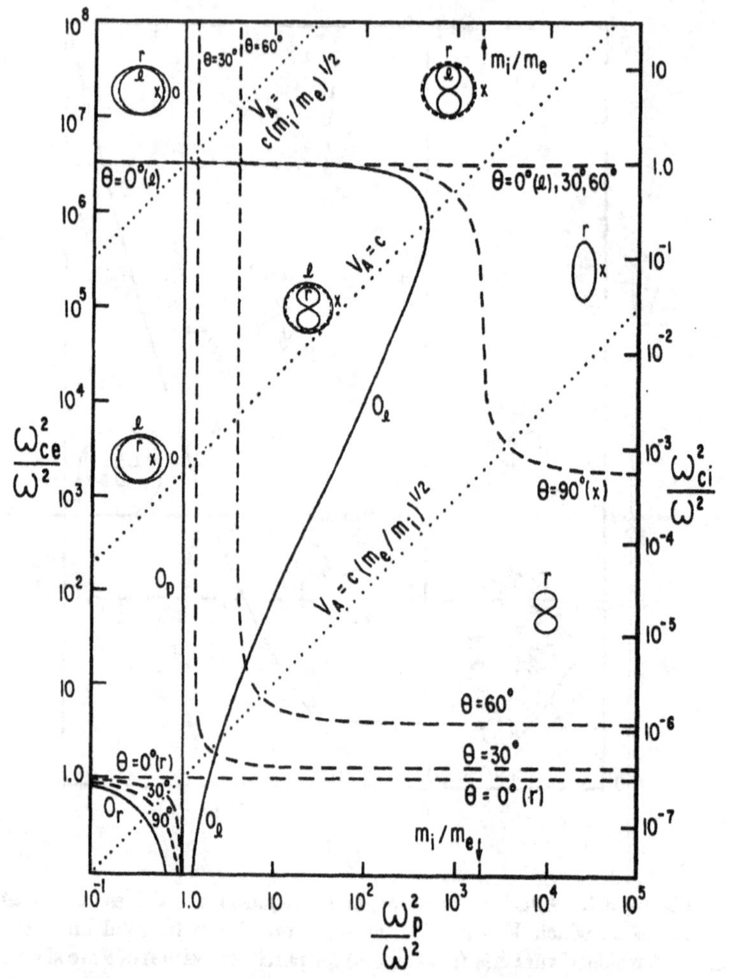

Figure 3.3. Allis diagram for low frequencies. Cut-offs and resonances are shown by solid and dashed lines, respectively, for an electron-proton gas ($m_i/m_e = 1836$). The dotted lines indicate particular ratios of the Alfvén speed, V_A, to c. The small diagrams depict $V(\theta)$ for the various modes of propagation.

gions where V^2 is negative no propagation is possible; these regions are sometimes called "stop bands."

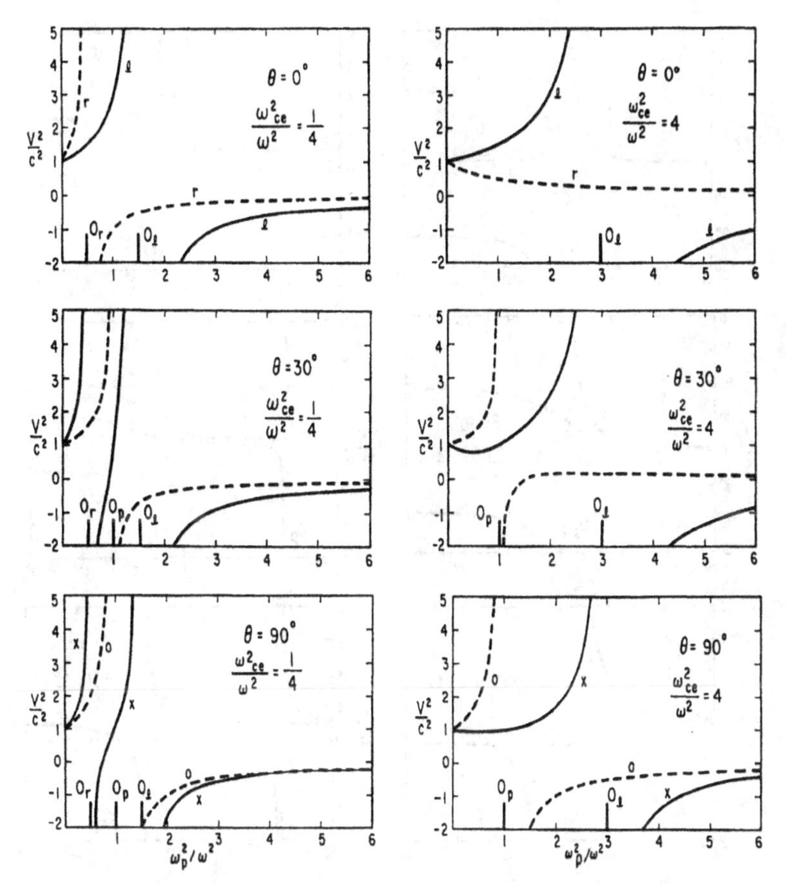

Figure 3.4. Phase velocity for high-frequency waves. The variation of V^2/c^2 with plasma density is shown for ω equal to half and twice ω_{ce}, respectively. The vertical lines at the bottom of each figure indicate the location of the various cut-offs.

A number of points may be noted in Figure 3.4. The ordinary wave always cuts off, of course, at ω_p equal to or greater than ω. This plasma cut-off is denoted by O_p in the figure. The extraordinary wave cuts off at densities greater than the left cut-off, O_l (or at frequencies less than ω_{ol}). For ω_{ce} less than ω the extraordinary wave also cuts off at densities

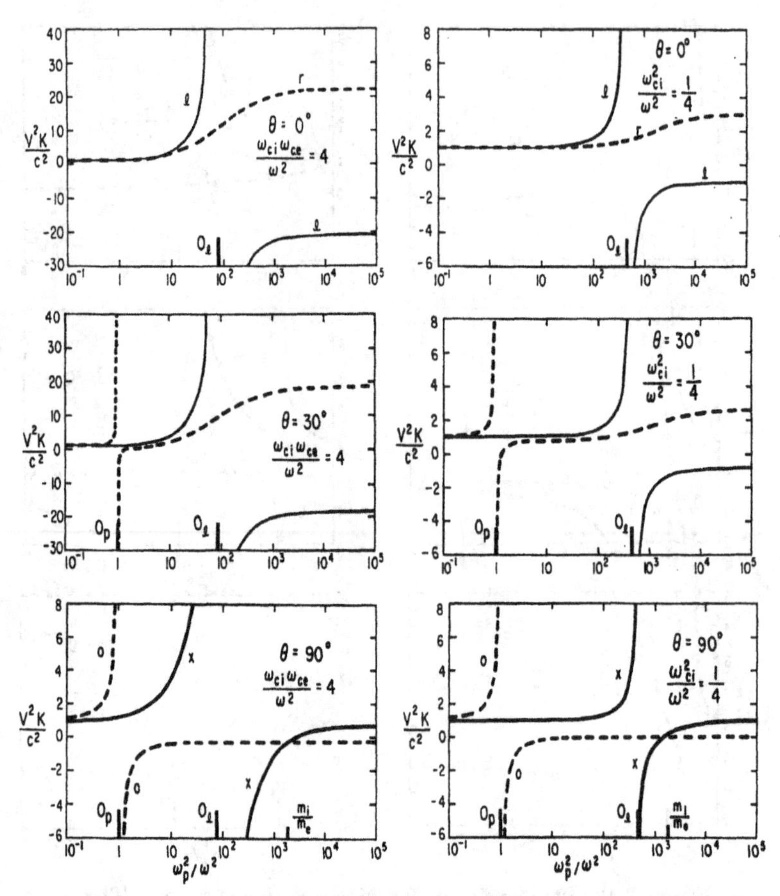

Figure 3.5. Phase velocity for low-frequency waves. The ratio of V^2 to the hydromagnetic value, c^2/K, is plotted for ω equal to $(\omega_{ce}\omega_{ci})^{\frac{1}{2}}/2$ and $2\omega_{ci}$, respectively. The vertical lines at the bottom of each diagram indicate the position of the plasma cut-off (O_p) and the left cut-off (O_l).

above the right cut-off, O_r; with increasing density, propagation becomes possible again between the upper hybrid resonance and the left cut-off. In this latter domain, at frequencies between ω_{h2} and ω_{ol}, the electric field is predominantly in the x direction, parallel to the propagation vector $\boldsymbol{\kappa}$; as ω_c/ω goes to zero this wave goes over into a purely electrostatic plasma wave. Thus the waves in this domain may be identified with electron waves, which, as we have already seen, were eliminated as an entirely separate mode by the neglect of the ∇p_e term in the generalized Ohm's Law. In the two lowest diagrams of Figure 3.5 we see that while the behaviour of the ordinary wave, across the magnetic field, is unaffected by high ω_{ce}/ω, and does not propagate at densities above plasma cut-off, the extraordinary wave, which disappears at the left cut-off, O_l, reappears at sufficiently high density, provided that the frequency is less than the lower hybrid frequency. For ω_{ci}/ω large the extraordinary wave becomes the magnetosonic wave discussed in the preceding section.

b. Propagation Parallel to **B**. We again consider a wave in the x direction, but assume that **B** in the undisturbed plasma has only the component B_x. Now the components E_x and j_x are uncoupled to the other wave components, yielding the familiar electron oscillations at the plasma frequency, ω_p. The components in the y and z direction yield two simultaneous equations for j_y and j_z. The condition that the determinant of the coefficients in these equations vanish yields the following dispersion relation

$$\frac{c^2}{V^2} = 1 - \frac{\omega_p{}^2}{\omega^2 - \omega_{ce}\omega_{ci} \pm \omega(\omega_{ce} - \omega_{ci})} \tag{3-44}$$

The plus sign in equation (3-44) corresponds to a circularly polarized wave in which the electric vector rotates in the left-hand direction, the direction of the magnetic field being taken as positive (i.e., $\boldsymbol{\kappa}$ and **B** in the same direction). This is the same sense in which the positive ion gyrates in a magnetic field. This wave is therefore called a left-handed wave, and denoted by the symbol l. The minus sign yields the right-handed, or r wave.

As in equation (3-34), we recover equation (3-7) when ω_{ce} and ω_{ci} vanish, and equation (3-28) for Alfvén waves when ω is very small. When V is much less than c, equation (3-44) yields

$$\frac{V^2}{V_A^2} = \left(1 \mp \frac{\omega}{\omega_{ci}}\right)\left(1 \pm \frac{\omega}{\omega_{ce}}\right) \tag{3-45}$$

where the upper and lower signs yield l and r waves, respectively.

The resonances for these waves, obtained very simply from equation (3-44), are

$$\omega = \begin{cases} \omega_{ci} & \text{for the } l \text{ wave} \\ \omega_{ce} & \text{for the } r \text{ wave} \end{cases} \tag{3-46}$$

It is evident physically that a resonance must arise when the electric vector rotates in the same sense and at the same frequency as a particle gyrates. These cyclotron resonances are shown as dashed lines on Figures 3.2 and 3.3. One may question why the extraordinary wave, travelling perpendicular to **B**, does not show a resonance at ω equal to ω_{ce}, since the oscillating electric field is entirely transverse to **B**. Computation of E_x and E_y in this situation shows that **E** is exactly circularly polarized in the plane perpendicular to **B**, and is rotating in the left-handed sense. Thus there is no accleration of individual electrons.

If V is set equal to infinity in equation (3-44), we may compute the cut-off frequencies. It follows readily that the l wave cuts off at ω_{ol}, the r wave at ω_{or}, where ω_{ol} and ω_{or} are defined in equations (3-40) and (3-41). It is for this reason that the two cyclotron cut-offs are designated as left and right cut-offs.

The behaviour of the r and l waves under different conditions is portrayed in the upper diagrams of Figures 3.4 and 3.5. For ω_{ce} less than ω the dispersion curves do not differ much from that for the ordinary wave travelling perpendicular to **B**, except that the cut-offs are displaced to each side of the plasma cut-off, at ω equal to ω_p. For ω_{ce} greater than ω, however, the r wave propagates for any value of ω_p^2/ω^2, as shown in the top diagrams of Figure 3.5, as well as in the right-hand diagram at the top of Figure 3.4. Similarly, the l wave propagates for any value of ω_p^2/ω^2 if ω_{ci} is greater than ω.

Evidently both l and r waves have stop bands when the appropriate ω_c/ω is less than unity and the density is sufficiently high. For low density, however, these stop bands become restricted to a narrow range of ω_c/ω slightly less than unity. If V_A is as great as $c(m_i/m_e)^{1/2}$ corresponding to a value of ω_{ce}^2/ω_p^2 exceeding m_i/m_e, and thus to relatively low plasma densities or high magnetic fields, the cut-off frequency ω_{0l} lies extremely close to the resonant frequency ω_{ci}, as may be seen from Figure 3.3, and the stop band for the l waves virtually disappears.

At high densities and at frequencies intermediate between ω_{ce} and ω_{ci} the r wave has an interesting physical interpretation. If we take ω_p^2/ω^2 to be large compared to m_i/m_e, equation (3-44), indicates that $(V/c)^2$ under these conditions equals $\omega\omega_{ce}/\omega_p^2$, or $\omega B/4\pi e n_e c$. Evidently this velocity is independent of electron or ion mass. The lines of force are helical and rotate, carrying the electrons with them; the ponderomotive force on the gyrating electrons is balanced directly by the magnetic stresses associated with the helical lines of force, and inertial forces are unimportant.

c. Propagation in an Arbitrary Direction. When θ is intermediate between 0 and $\pi/2$ the dispersion relation becomes rather complicated, and will not be given here. For high frequencies, when ω_{ci}/ω may be neglected, the relation between V and θ is well known from ionospheric studies, and is discussed by Mitra (30) and Ratcliffe (33). The equations in the more general case have been given by Aström (5) and by Allis and his colleagues (4). Here only certain general results will be given.

One very important result is that the cut-off frequencies are independent of θ. This conclusion may be established if the dispersion relation is expressed as a quadratic equation in c^2/V^2. The term independent of c^2/V^2 in this equation turns out to be independent of θ, and hence the values of ω for which this term vanishes, and one root for V^2 is therefore infinite, are also independent of θ.

The resonances, however, do depend on θ. The angle θ at which V vanishes, which we call a "resonant angle," is given by

$$\sin^2 \theta = \left(1 - \frac{\omega^2}{\omega_p^2}\right)\left\{1 - \frac{\omega^2(\omega^2 - \omega_{ce}\omega_{ci})}{(\omega^2 - \omega_{ce}^2)(\omega^2 - \omega_{ci}^2)}\right\}^{-1} \quad (3\text{-}47)$$

In general the relationship between ω_p^2/ω^2 and ω_{ce}^2/ω^2 given by this equation, for arbitrary θ, is intermediate between those for θ equal to 0° and for θ equal to 90°. The resonance lines for θ equal to 30° are shown in Figures 3.2 and 3.3.

Values of V^2/c^2 for θ equal to 30° are shown in Figures 3.4 and 3.5 for the same parameters as for the other two directions of propagation. To provide additional qualitative information on how the phase velocity $V(\theta)$ varies with the direction of propagation, qualitative polar plots of $V(\theta)$ are given in each region of the Allis diagrams in Figures 3.2 and 3.3. Imaginary values of V are not shown. The direction $\theta = 0°$ is taken to be vertical on the diagram, and the wave normal surfaces are symmetrical about this vertical direction. The modes for 0° and 90° are indicated by the letters r, l, o, and x. To indicate the scale of the velocities, a dashed circle indicates c, the velocity of light on Figure 3.2. In some of the $V(\theta)$ plots in Figure 3.3 the velocities are much less than c, and the dashed circle is omitted from these plots.

Across a cut-off line an entire mode appears or disappears, with V approaching infinity on one side. Across a resonance line the velocity in a particular direction vanishes, and between the resonance lines for $\theta = 0°$ and $\theta = 90°$ one mode disappears by a gradual widening of the resonance angle.

When the r wave and the x wave lie on the same wave normal surface, as in the lower left-hand corner of Figure 3.2, the r and x modes are said to correspond. Examination of Figures 3.2 and 3.3 shows that the correspondence between the r and l modes on the one hand and the o and x modes on the other is reversed across the plasma cut-off line. The correspondence also reverses across the line ω^2 equal to $\omega_{ci}^2 + \omega_p^2 m_e/m_i$, where the o and x modes cross, as in the two bottom diagrams of Figure 3.5.

The propagation of r and l waves at high densities, if ω is less than the appropriate ω_c, has already been commented upon. The variation of the resonance line for r waves with θ shown on the lower right-hand side of Figure 3.3 indicates that for ω slightly less than ω_{ce}, this propagation is possible only in a relatively narrow cone, which broadens out as ω falls well below ω_{ce}. From equation (3-47) we see that for large ω_p^2/ω^2, the resonance angle broadens out to 30° for ω_{ce}^2/ω^2 equal to 4/3. This mode of propagation has been observed for waves along the earth's magnetic field out to several earth radii (the so-called "whistler mode").

As ω decreases below ω_{ci} the resonance cone for l or o waves also broadens out, but very much more rapidly. It may be seen from equation (3-47) that for large ω_p^2/ω^2, θ increases to 30° as ω_{ci}^2/ω^2 increases from 1 to $1 + m_e/3m_i$, or about 1.00018. For ω_p/ω greater than sec θ the resonance line at an arbitrary θ is indistinguishable from the resonance line for θ equal to zero, as shown by the dashed line for $\theta = 30°$ at the top of Figure 3.3. The velocity of these left-handed waves varies about as sin θ, as we have seen already in Figure 3.1. This mode of plasma oscillation, at frequencies somewhat less than the ion cyclotron frequency, has been called an "ion cyclotron wave." Such waves have been extensively analyzed by Stix (35) in the particular case of axially symmetrical waves.

3.5 Damping and Excitation of Waves

In the hypothetical medium assumed in the earlier sections of this chapter the different kinds of infinitesimal waves propagate indefinitely without loss of energy. In a conventional fluid damping is produced by collisions, while excitation is usually produced by disturbances at some boundary. These same processes can also occur, of course, in plasmas. Viscosity, electrical resistivity, and thermal conductivity can all lead to conversion of wave energy into heat within an ionized gas. Excitation of waves at plasma boundaries can also occur. These processes are straightforward and well known. Here we shall be concerned with processes occurring in the volume of the gas in the absence of collisions. As we shall see, interactions between waves and particles moving through the gas can decrease the energy of a wave, or, if the velocity distribution is non-Maxwellian, wave amplification may result. Processes which produce exponential wave growth in the presence of a non-Maxwellian velocity distribution are known as "micro-instabilities," since their analysis depends on the microscopic description of the gas. A detailed consideration of these processes requires extended use of the Boltzmann equation, which is beyond the scope of this book. A detailed treatment of some of these problems has recently been given by Stix (37). Here we discuss the basic physical principles underlying three of the most important mechanisms for wave damping or excitation.

a. Landau Damping. It was pointed out by Landau (27) that in accordance with the linearized theory a longitudinal wave would be damped by particles moving at nearly the phase velocity of the wave, if the velocity distribution is assumed Maxwellian. When the phase velocity, V, of the wave much exceeds the root mean square particle velocity, this damping is slow and is produced by the acceleration of particles moving less rapidly than the wave. Particles with a velocity exceeding V are decelerated, but if $df^{(0)}/dw$, the derivative of the initial velocity distribution function, is negative for w equal to V, the net effect will be damping of the wave, whose amplitude will then vary as $\exp(-\sigma t)$.

The dynamical details of this process have been considered by Dawson (14), who computes trajectories of single particles. The wave is assumed to start at the time $t = 0$, when the velocity distribution is Maxwellian. At first, all particles will start to exchange energy with the wave. As time goes on, however, the particles whose initial velocity, u, relative to the wave is appreciable will reach a state where they alternately take energy from and return energy to the wave as they pass through the crests and troughs of the wave. Particles with a smaller absolute value of u take longer to reach this state, and as time goes on more and more energy appears in particles of lesser u^2. If the electric field is sufficiently great, this successive acceleration of slower and slower particles will terminate because of particle trapping, a process which we now consider in more detail.

If we view a travelling longitudinal wave in a reference frame moving at the velocity V, the electric potential will be independent of time; if for the moment we set σ equal to zero, we may write

$$U = U_0 \cos \kappa x \tag{3-48}$$

If the particle has a velocity u in the x direction, relative to the wave, and is at the position x, it will be trapped in a potential minimum if

$$\frac{1}{2} mu^2 + \frac{ZeU_0}{c} \cos \kappa x < \left| \frac{ZeU_0}{c} \right| \tag{3-49}$$

When such a trapped particle is reflected or turned around by the wave its energy in the wave frame is unaltered, but in the plasma frame its kinetic energy is altered by an amount $2mVu$. If we compare the initial energy in the plasma frame with the mean energy in this frame over a long period, averaging over many reflections of the particle, back and forth relative to the wave, the excess energy given up by a trapped particle is mVu. This energy will be given up to the wave by all particles for which u and x satisfy inequality (3-49). If we define u_m as the root mean square value of the maximum u at which trapping is possible, replacing the inequality by an equal sign and averaging over all x, we find

$$u_m{}^2 = \left| \frac{2ZeU_0}{mc} \right| \tag{3-50}$$

For u appreciably less than u_m, and for particles located not too far from the troughs of the wave potential, the particles will execute simple harmonic motion in these troughs. The frequency of oscillation for these trapped particles, which we denote by ω_t, is readily obtained from equation (3-48) on the assumption that κx is small, yielding

$$\omega_t = \kappa \left| \frac{ZeU_0}{mc} \right|^{1/2} \tag{3-51}$$

The time required for particles of the appropriate energy to become trapped is of order $1/\omega_t$. Evidently, if σ/ω_t is small compared to unity, where σ is again the decay rate of U_0, U_0 will be nearly constant during the trapping time and particle trapping will in fact occur. If σ/ω_t is large, however, the decay of U_0 during the time $1/\omega_t$ is large, and the wave damps out so rapidly that particle trapping is unimportant.

These conclusions have direct relevance to the phenomenon of Landau damping. We first consider the case where σ/ω_t is small and particle trapping occurs. In this case the transfer of energy from the wave to particles with small u must cease when t much exceeds $1/\omega_t$. Once trapping is complete, the wave cannot accelerate particles any longer, and the wave will propagate without change of amplitude.

The magnitude of Landau damping may be computed approximately from the condition that the energy lost by the wave during the trapping time equals the energy gained by the trapped particles. Following the analysis by Jackson (25) we may compute the total energy, W, gained by the trapped particles in the form

$$W = -\int_{-u_m}^{+u_m} nf^{(0)}(w)dw \cdot mVu = -nmVu_m{}^3 \frac{2}{3} \frac{df^{(0)}(V)}{dw} \tag{3-52}$$

where, as before, u is defined as $w - V$, and mVu is the energy given up by each particle when it becomes trapped. In this equation we have assumed that $f^{(0)}(w)$ is normalized to give unity on integration over w. Averaging results over space, we may replace u_m^3 by the average value found from equation (3-50). The time during which this energy W is transferred from the particles to the wave will be approximately $1/\omega_t$, and the mean rate of power transfer during this time will be $W\omega_t$. If we divide this rate by $\kappa^2 U_0^2/8\pi c^2$, the density of potential and kinetic energy in the wave, the quotient is 2σ; the amplitude decay rate is σ. In the case of electrons moving through a plasma in which an electron wave is propagating, Z equals -1, V equals ω_p/κ, and we obtain

$$\sigma = b \frac{\omega_p}{(\kappa h)^3} e^{-0.5/(\kappa h)^2} \tag{3-53}$$

where equations (2-3) and (3-8) have been used for the Debye shielding distance, h, and the plasma frequency, ω_p, respectively. The numerical constant, b, is about $4/3\pi^{1/2}$ on the basis of the above crude analysis; Landau (27) gives equation (3-53) with b equal to $(\pi/8)^{1/2}$. Thus, when $\sigma\omega_t$ is small, the energy taken out of the wave by Landau damping during the time $1/\omega_t$ is mostly transferred to the trapped particles; for times larger than $1/\omega_t$, Landau damping ceases in a collisionless gas.

In the converse case, when σ/ω_t is large, particle trapping is unimportant. Equation (3-53) is still valid, however, and the damping rate is now independent of time. It is evident from equation (3-51) that σ/ω_t will be large for waves of sufficiently small amplitude.

It will be noted that if we assume a different set of initial conditions when the wave is first assumed to propagate, these results are altered. In particular, if we assume that $f(V + u)$ equals $f(V - u)$ for a substantial range in u, Landau damping can be eliminated entirely, even for waves of infinitesimal amplitude, and, if no collisions occur, wave propagation in a completely steady state becomes possible (8) regardless of amplitude.

Finally all this analysis is applicable only if

$$(\kappa h)^2 = \frac{kT}{mV^2} \ll 1 \tag{3-54}$$

When this condition is violated the phase velocity is comparable with or less than the random particle velocities, the damping becomes strong and the assumptions made above are no longer realistic. The wavelength is then comparable to or less than the Debye shielding distance, the formulae for damping given by Landau take quite a different form from equation (3-53); σ is now given by

$$\sigma \approx \kappa \left(\frac{kT}{m}\right)^{1/2} \tag{3-55}$$

The physical phenomenon occurring is quite different from that discussed above. For wavelengths much less than h, the long range forces do not suffice to preserve collective behaviour. The particles move in straight lines, unaffected by electrostatic forces, and any macroscopic density fluctuations are dissipated in about the time required for random thermal motions to carry a particle across a wavelength.

b. Cyclotron Damping. A powerful form of damping appears when some particles experience a perturbing force which oscillates at their cyclotron frequency. This can readily occur, for example, if a wave with an electric vector transverse to **B** is propagating at least partly in the direction of **B**. Then to a particle moving along **B** the frequency of the oscillating electric field will be Doppler shifted, and at some velocity will be nearly equal to the cyclotron frequency.

We compute the attenuation distance, d, resulting from cyclotron damping in a simple hypothetical case. A circularly polarized wave is assumed to propagate in the z direction, parallel to the magnetic field, with a phase velocity V. Particles moving with a velocity $w_{\parallel}$ along the magnetic field are assumed to have a Maxwellian velocity distribution at the plane $z = 0$. We compute the energy absorbed by the particles moving along the lines of force on the assumption that the damping is small and may be neglected in a preliminary analysis of the particle accelerations. For a positive particle we must consider a left-handed wave (if the propagation is in the same direction as **B**); the electric field is then given by

$$E_x = E_0 \cos(\kappa z - \omega t)$$
$$E_y = E_0 \sin(\kappa z - \omega t) \tag{3-56}$$

The phase velocity ω/κ for this wave is given in equation (3-44). The particle velocity, $w_{\parallel}$ in the z direction is unaffected by the electric field. We let $\Delta w_{\perp}$ be the increase in transverse velocity at a time t after crossing the plane $z = 0$. If the distribution of transverse velocities is initially isotropic (i.e., if particles have a random distribution of phases in their gyration about the magnetic field when $z = 0$), then by substituting equation (3-56) into the equation of motion (1-1), we readily obtain

$$\frac{m}{2}\Delta w_{\perp}^2 = \frac{2Z^2e^2E_0^2}{mc^2}\frac{\sin^2(\kappa w_{\parallel} + \omega_c - \omega)t/2}{(\kappa w_{\parallel} + \omega_c - \omega)^2} \tag{3-57}$$

Since t equals $z/w_{\parallel}$, equation (3-57) gives the increased energy of a group of particles as a function of distance, z, along the magnetic field. The decrease in energy flux of the wave, with increasing z, must just equal the increase in energy flux of all the particles. To compute this increase we multiply equation (3-57) by $w_{\parallel}n(w_{\parallel})dw_{\parallel}$, the flux of particles per cm² per sec whose velocity along the magnetic field lies between $w_{\parallel}$ and $w_{\parallel} + dw_{\parallel}$, and then integrate over $w_{\parallel}$ from zero to infinity. Since the velocity distribution is assumed initially Maxwellian, and since $w_{\parallel}$ does not change with time, $n(w_{\parallel})$ is the usual Maxwellian function. For large z the integrand peaks sharply at $w_{\parallel}$ equal to the "resonant velocity," w_r, given by

$$w_r = \frac{\omega - \omega_c}{\kappa} \tag{3-58}$$

Hence we may evaluate $w_{\parallel}n(w_{\parallel})$ at $w_{\parallel} = w_r$, and then integrate the remaining expression, obtaining a result proportional to z. This increase of the particle energy flux per unit distance may then be equated to the corresponding decrease of the Poynting flux $E_0^2/4\pi V$. Introducing the attenuation distance,

d, defined in equation (3-5), we obtain after some algebra

$$d = \frac{2}{\pi}\left(\frac{c}{V}\right)^2 \frac{\omega w_r}{\omega_{pr}^2} \tag{3-59}$$

where $(c/V)^2$ is given in equation (3-44) and where

$$\omega_{pr}^2(w_r) = \frac{4\pi w_r n(w_r) e^2}{m} \tag{3-60}$$

Evidently ω_{pr} is similar to the plasma frequency, but with $w_r n(w_r)$ replacing n, the total number of particles per cm^3, and with the mass of the particle under consideration replacing m_e. For electron cyclotron damping ω_{ce} and m_e appear in equations (3-58) and (3-60), respectively, and the sign of E_y in equation (3-56) must be changed to give a right-handed wave. From these equations it is readily shown that cyclotron damping may convert wave energy into particle energy in a relatively short distance.

In the derivation of equation (3-59), the change of E_0 with d was neglected, and the wave amplitude was taken to be infinitesimal. The first approximation does not affect the results, since if we take E_0 proportional to $\exp(-z/d)$, equation (3-59) can again be obtained, provided that d is large compared to the wavelength. The effect of finite amplitude has been considered in the more general analysis by Stix (37); as in the case of Landau damping, cyclotron damping will be limited by particle trapping for waves of sufficient amplitude.

c. Excitation, Two-Stream Instability. The two damping processes analyzed above will convert wave energy into random kinetic energy if the initial velocity distribution is reasonably close to Maxwellian. If the initial velocity distribution is sufficiently far from Maxwellian, the reverse process becomes possible and wave amplification can occur. A number of different mechanisms can operate to produce this effect.

The simplest amplification process is the converse of the Landau damping due to trapping of particles moving at the phase velocity of the wave. It is evident from the analysis given above—see equation (3-52)—that if $df^0(V)/dv$ is positive instead of negative, particle trapping will add to the wave energy rather than subtract from it.

A more rapid type of amplification is possible when one or more streams of electrons is traversing a plasma. Longitudinal electron waves can then be unstable, with rapid growth resulting.

The physical mechanism permitting the instability is that an electron beam travelling through an exponentially growing longitudinal wave tends to slow down relative to the wave. Let us consider a wave in a frame of reference where the wave velocity is zero. If the wave amplitude grows exponentially, then E_x may be written

$$E_x = Ae^{\sigma t} \sin \kappa x \tag{3-61}$$

The linearized equation of motion of the electron stream may be written

$$\frac{\partial u^{(1)}}{\partial t} + u \frac{\partial u^{(1)}}{\partial x} = -\frac{eE_x}{mc} \tag{3-62}$$

where u is the initial velocity of the stream and $u^{(1)}$ is the linearized perturbation in velocity. The corresponding equation for $n^{(1)}$, the perturbation in electron density, is

$$\frac{\partial n^{(1)}}{\partial t} + u \frac{\partial n^{(1)}}{\partial x} + n \frac{\partial u^{(1)}}{\partial x} = 0 \tag{3-63}$$

where n denotes the unperturbed density. Solution of these equations for $n^{(1)}$ yields a component of $n^{(1)}$ which is proportional to σ and which varies as $\sin \kappa x$, in phase with E_x. Because of this component the mean value of $n^{(1)}E_x$ differs from zero, and yields a retardation of the beam proportional to A^2. This retardation of the beam provides the energy available to increase the wave energy. Conversely, if the wave amplitude is decreasing the beam increases its velocity relative to the wave. If σ is entirely imaginary, $n^{(1)}$ is out of phase with E_x, and there is no net change in the energy of the beam over one wavelength.

To derive explicitly the conditions under which wave amplification is possible, we must consider at least two groups of particles with particle densities, charges, masses, and velocities given by n_j, Z_je/c, m_j and u_j, respectively. Equations (3-62) and (3-63) then hold separately for each $u_j^{(1)}$ and $n_j^{(1)}$. To determine E_x we may use equation (3-2), with j_x determined by the linearized relationship

$$j_x = \sum_j \frac{Z_je}{c} (n_j^{(1)}u_j + n_ju_j^{(1)}) \tag{3-64}$$

If equations (3-62) and (3-63) are substituted into equation (3-2), with all quantities assumed proportional to exp $i(\kappa x - \omega t)$, we obtain

$$\sum_j \frac{\omega_{pj}^2}{(\omega - \kappa u_j)^2} = 1 \tag{3-65}$$

where ω_{pj}^2 is given in the familiar equation (3-8), with use of the proper value of n_j, Z_j, and m_j for the jth component. Equation (3-65) was first derived by Haeff (19, 20) and Pierce (32). A survey of the field, with recent references, is given by Bernstein and Trehan (9).

In the simple case that two electron streams are present with equal and opposite velocities, $\pm u/2$, and with equal particle densities, equation (3-65) is a simple quadratic equation in ω^2, that may be solved explicitly, yielding (26)

$$\omega^2 = \omega_p^2 + \frac{\kappa^2 u^2}{4} \pm (\omega_p^4 + \omega_p^2\kappa^2u^2)^{1/2} \tag{3-66}$$

The frequency ω_p is the plasma frequency for either of the streams separately. It is readily seen that ω^2 has a negative root if

$$\kappa u < 2^{3/2}\omega_p \tag{3-67}$$

Evidently if each stream travels a wavelength in a time much less than $1/\omega_p$, the disturbance of that wavelength is stable. The waves of imaginary ω are standing waves which grow or decay exponentially; as seen in a reference frame travelling with one of the streams these waves have a frequency $\kappa u/2$. If $\kappa u/\omega_p$ is small the rate of growth is given by

$$\sigma \equiv i\omega \approx \pm \frac{\kappa u}{2} \tag{3-68}$$

These long waves, which relative to the particle streams have a very low frequency, have no counterpart in the absence of streaming motion; they have a slow growth (or decay) rate.

The maximum growth rate is found for $\kappa u/\omega_p$ equal to $3^{1/2}$, at which point σ equals $\omega_p/2$. As seen by one of the streams the apparent real frequency is 0.86 times the plasma frequency for that stream. The growth rate is evidently very rapid. In general the two-stream instability grows most rapidly when the streams see a frequency relatively close to their proper plasma frequencies. In this case, particle bunching produced by the wave increases the wave energy at the greatest rate.

The treatment for two general streams, with differing particle densities or masses, is somewhat more complicated. However, the region of instability and the maximum growth rates may be obtained analytically (9). We choose subscripts 1 and 2 so that ω_{p1}^2 exceeds ω_{p2}^2, and define ζ as the ratio $\omega_{p2}^2/\omega_{p1}^2$. For convenience we view the wave from a reference frame in which the drift velocity of beam 1, with the greater plasma frequency, vanishes; again, let u denote the relative velocity of the two beams. The ratio ω/ω_{p1} is now a function of the two variables, ζ and κu. Instability arises if

$$\frac{\kappa u}{\omega_{p1}} < (1 + \zeta^{1/3})^{3/2} \tag{3-69}$$

This result reduces to equation (3-67) when ζ is unity. The maximum growth rate is reached for $\kappa u/\omega_{p1}$ between unity and the upper limit in equation (3-69); for small ζ this maximum σ is given by

$$\frac{\sigma}{\omega_{p1}} = \frac{3^{1/2}}{2}\left(\frac{\zeta}{2}\right)^{1/3} \tag{3-70}$$

a relation derived by Buneman (11). Thus if a stream of electrons, with density n_2, is moving through a plasma, of electron density n_1, the maximum growth rate is about $0.7(n_2/n_1)^{1/3}\omega_{p1}$. The real part of ω is slightly less than κu, corresponding to the physical condition that the drift velocity u must exceed ω/κ, the phase velocity of the waves.

The same analysis applies if the electrons and positive ions have a relative drift velocity u. In this case ζ equals Zm_e/m_i and for an electron-proton gas the maximum σ is 0.056 ω_p; this maximum growth rate occurs for disturbances of wave number about equal to ω_p/u. In accordance with the definitions above, the wave is viewed in the reference frame of the electrons. In the reference frame of the positive ions the frequency differs by an amount κu. The real part of ω, as measured in this frame, is at most equal to σ, and is thus substantially less than ω_p; the positive ions participate in the oscillations, and their greater mass slows down both the oscillations and their rate of growth.

When a continuous distribution of velocities is considered the analysis becomes more involved. Instabilities arising from net drifts of ions relative to electrons have recently been considered by various authors (11, 24) on the assumption that ions and electrons each have Maxwellian velocity distributions, with a relative drift u. For instability u must not only be less than a critical upper value, as in equation (3-69), but must also exceed a lower value. If the electron and ion temperatures, T_e and T_i, are about equal, this lower critical velocity is about equal to the random electron velocity. For T_i/T_e less than 0.1 this critical velocity is about equal to the velocity of acoustic or positive-ion waves; according to equation (3-21) this velocity equals $(m_e/m_i)^{1/2}$ times the random electron velocity.

If a magnetic field is assumed in the unperturbed state, amplification of transverse electromagnetic waves can also occur, a subject first extensively analyzed by Bailey (7); the dispersion relations for a number of simple cases have been considered by Bernstein and Trehan (9). As with longitudinal waves, relatively rapid wave amplification can occur under a variety of circumstances.

Other sources of wave excitation are also possible in a plasma when a magnetic field is present. As shown in the next chapter, a plasma can be unstable against certain types of hydromagnetic disturbances if the pressure is sufficiently anisotropic. Sagdeyev and Shafranov (34) have shown that even if $p_\perp$ and $p_\parallel$ differ only slightly there will be a slow amplification of electromagnetic waves. This instability is produced by particles moving along the lines of force at the resonant velocity w_r, defined in equation (3-58); for these particles the Doppler-shifted wave frequency equals the cyclotron frequency. A related instability for nearly longitudinal waves, at a frequency about equal to ω_c, has been studied by Harris (21).

References

1. Alexeff, I. and R. V. Neidigh, *Phys. Rev. Letters*, **7**, 223 (1961).
2. Alfvén, H., *Ark. Mat., Astr. Fysik*, **29B**, No. 2 (1942).
3. Alfvén, H., *Cosmical Electrodynamics*, Chapter IV, Clarendon Press, Oxford, 1950.
4. Allis, W. P., S. J. Buchsbaum, and A. Bers, *Wave Propagation in Anisotropic Plasma*, in press.
5. Aström, E., *Ark. Fys.*, **2**, 443 (1950).
6. Auer, P. L., H. Hurwitz, and R. D. Miller, *Phys. Fluids*, **1**, 501 (1958).
7. Bailey, V. A., *Australian J. Sci. and Industrial Res.*, **1**, 351 (1948).
8. Bernstein, I. B., J. M. Greene, and M. D. Kruskal, *Phys. Rev.*, **108**, 546 (1957).
9. Bernstein, I. B. and S. K. Trehan, *Nuclear Fusion*, **1**, 3 (1960).
10. Bohm, D. and E. P. Gross, *Phys. Rev.*, **75**, 1851 (1949); 1864 (1949).
11. Buneman, O., *Phys. Rev.*, **115**, 503 (1959).
12. Clemmow, P. C. and R. F. Mullaly, *The Physics of the Ionosphere*, Physical Society, London, 1955, p. 340.
13. Cowling, T. G., *Magnetohydrodynamics*, Interscience, New York, 1957.
14. Dawson, J., *Phys. Fluids*, **4**, 869 (1961).
15. Ferraro, V. C. A., *Proc. Roy. Soc.* (*London*), **A233**, 310 (1955).
16. Field, G. B., *Astrophys. J.*, **124**, 555 (1956).
17. Fried, B. D. and R. W. Gould, *Phys. Fluids*, **4**, 139 (1961).
18. Gabor, L., *Brit. J. Appl. Phys.*, **2**, 209 (1951).

19. Haeff, A. V., *Phys. Rev.*, 74, 1532 (1948).
20. Haeff, A. V., *Proc. Inst. Radio Engrs.*, 37, 4 (1949).
21. Harris, E. G., *J. Nuclear Energy, Part C, Plasma Phys.*, 2, 138 (1961).
22. Herlofson, N., *Nature*, 165, 1020 (1950).
23. Hernqvist, K. G., *J. Appl. Phys.*, 26, 1029 (1955).
24. Jackson, E. A., *Phys. Fluids*, 3, 786 (1960).
25. Jackson, J. D., *J. Nuclear Energy, Part C, Plasma Phys.*, 1, 171 (1960).
26. Kahn, F. D., *J. Fluid Mech.*, 2, 601 (1957).
27. Landau, L., *J. Phys. (U. S. S. R.)*, 10, 25 (1946).
28. Looney, D. H. and S. C. Brown, *Phys. Rev.*, 93, 965 (1954).
29. Lundquist, S., *Ark. Fys.*, 5, 297 (1952).
30. Mitra, S. K., *The Upper Atmosphere*, Royal Asiatic Society of Bengal, 1947, p. 143.
31. Penning, F. M., *Physica*, 6, 241 (1926).
32. Pierce, J. R., *J. Appl. Phys.*, 19, 231 (1948).
33. Ratcliffe, J. A., *The Magneto-Ionic Theory and its Application to the Ionosphere*, Cambridge University Press, Cambridge, 1959.
34. Sagdeyev, R. S. and V. Shafranov, *J. Exptl. Theoret. Phys. (U. S. S. R.)*, 39, 181 (1960); *Soviet Phys. JETP*, 12, 130 (1961).
35. Stix, T., *Phys. Rev.*, 106, 1146 (1957).
36. Stix, T., *Phys. Fluids*, 3, 19 (1960).
37. Stix, T., *Theory of Plasma Waves*, McGraw-Hill, New York, 1962.
38. Tidman, D. A., *Phys. Rev.*, 117, 366 (1960).
39. Tonks, L. and Langmuir, I., *Phys. Rev.*, 33, 195 (1929).
40. van de Hulst, H. C., *Problems of Cosmical Aerodynamics*, Central Air Documents Office, Dayton, Ohio, 1951, Chapter VI.

CHAPTER 4

Equilibria and Their Stability

When all the macroscopic quantities characterizing a plasma are constant in time, the plasma is said to be in equilibrium. In the laboratory and in astrophysics particular interest attaches to a "confined" plasma, which is held in a steady state within a finite region, surrounded by a magnetic field. Because of their possible importance in the controlled release of thermonuclear power, confined plasmas have received much attention during the last few years and many possible magnetic configurations have been analyzed.

The existence and nature of equilibrium states of a plasma and the extent to which these states are stable can be analyzed with the macroscopic equations given in Chapter 2, without recourse to the more powerful but vastly more complex treatment involving the Boltzmann equation and the velocity distribution function. We shall analyze here the equilibria expected in relatively simple geometries, and shall deal very briefly with the stability problem in order to indicate the type of instabilities that may arise in an ionized gas. The first section discusses the principles involved, while subsequent sections apply these principles to various plasma configurations.

4.1 Principles of Stable Equilibrium

For a system to persist in a particular state it must satisfy two conditions: (a) it must be in equilibrium, and (b) the equilibrium must be stable. A fully ionized gas will be in equilibrium if it satisfies equations (2-11) through (2-19) with all the time derivatives set equal to zero. It will be stable if all types of infinitesimal perturbations lead to damped oscillations about this equilibrium state, and unstable if one or more types of perturbation grow exponentially. We examine each of these two conditions in more detail.

The analysis of any equilibrium state is straightforward. If we neglect terms of order m_e/m_i, equations (2-20) and (2-21) are the ones that must be satisfied. A confined plasma is possible in equilibrium only if the resistivity, η, vanishes; alternatively one might consider a steady state in which the plasma slowly diffuses outwards as a result of finite resistivity, while some source of new gas maintains the density uniform. We shall here neglect the $\eta\mathbf{j}$ term in equation (2-21), in which case this equation gives a relationship between $\mathbf{E}$ and $\mathbf{v}$ but is not required for determining p, $\mathbf{j}$ and $\mathbf{B}$.

The basic relationship which must now be satisfied in equilibrium is equation (2-20). If we neglect the gravitational potential, ϕ, we obtain

$$\nabla p = \mathbf{j} \times \mathbf{B} = \frac{1}{4\pi}\mathbf{B}\cdot\nabla\mathbf{B} - \nabla\frac{B^2}{8\pi} \qquad (4\text{-}1)$$

where we have eliminated $\mathbf{j}$ by means of equation (2-19), with $\partial\mathbf{E}/\partial t$ set equal to zero. Equation (4-1) is sometimes called the "magnetostatic equation." The term in $\mathbf{B}\cdot\nabla\mathbf{B}$ on the right-hand side of equation (4-1) represents the stresses due to tensions along the lines of force; when the lines are curved, volume forces can appear. The term in ∇B^2 represents stresses due to mutual repulsion of the lines of force; these stresses are similar in effect to a pressure tensor isotropic in the plane perpendicular to $\mathbf{B}$.

Several simple results may be derived immediately. If we take the scalar product of equation (4-1) first with $\mathbf{B}$ and then with $\mathbf{j}$, the right-hand side of the equation vanishes in each case. Hence ∇p can have no component parallel either to $\mathbf{B}$ or to $\mathbf{j}$; both $\mathbf{B}$ and $\mathbf{j}$ must be parallel to the isobaric surfaces. If inertial forces are important or if the pressure tensor is anisotropic, equation (4-1) becomes substantially more complicated and these simple results no longer hold.

Once p and $\mathbf{B}$ are known for the equilibrium situation, the analysis of infinitesimal perturbations is also straightforward, and is in principle identical with the analysis of infinitesimal waves given in the previous chapter. In Chapter 3 the equilibrium state was uniform, and thus all coefficients in the differential equations for the perturbed quantities were constant. In the present instance these coefficients are functions of position, and analysis of the perturbations is an intricate eigenvalue problem. The solution of such a problem gives all the normal modes, and if any of these modes increases exponentially instead of oscillating the equilibrium is unstable.

Here we shall discuss the stability in a much simpler manner, by considering the change of the potential energy for an arbitrary deformation. The potential energy, W, of the plasma may be written

$$W = \int \left\{ \frac{B^2}{8\pi} + \frac{3p}{2} + \rho\phi \right\} d\tau \qquad (4\text{-}2)$$

where $d\tau$ is a volume interval, and the integral extends over the entire region occupied by the plasma and the surrounding vacuum; this volume will be finite only if the system is enclosed by a perfect conductor. Since the resistivity and viscosity are neglected, the system is nondissipative and the total energy, given by the sum of W and the kinetic energy, is constant.

Let an equilibrium system be perturbed by imposing an arbitrary displacement, $\boldsymbol{\xi}$, a function of initial position. To first order in $\boldsymbol{\xi}$ the change δW in the potential energy must vanish, since this is the condition for an equilibrium state. The stability or instability of the equilibrium is determined by the sign of $\delta W(\boldsymbol{\xi}, \boldsymbol{\xi})$, the value of δW when all terms of order ξ^2 are retained. If $\delta W(\boldsymbol{\xi}, \boldsymbol{\xi})$ is positive, the kinetic energy cannot exceed the initial value of δW, and the perturbation will not grow. If $\delta W(\boldsymbol{\xi},\boldsymbol{\xi})$ is negative, however, $|\delta W|$ and the kinetic energy can grow together as ξ^2 increases, and the system is unstable.

More quantitatively, we may write the conservation of

energy in the form

$$\frac{1}{2}\int \rho \left(\frac{d\xi}{dt}\right)^2 d\tau + \delta W(\xi, \xi) = 0. \tag{4-3}$$

If we assume that ξ varies as exp $(-i\omega t)$, we obtain

$$\omega^2 = \frac{\delta W(\xi, \xi)}{\frac{1}{2}\int \rho\xi^2 d\tau}. \tag{4-4}$$

Thus if δW is negative, ω is imaginary, and the perturbation grows exponentially. Only if ξ is an eigenfunction of the problem will the perturbation increase at the same rate throughout the plasma. Even with an approximate ξ equation (4-4) gives correctly the general magnitude of the rate at which ξ increases.

The general equations for $\delta W(\xi, \xi)$ have been derived from the macroscopic equations by Bernstein, Frieman, Kruskal, and Kulsrud (2), who have established as a necessary and sufficient condition for instability that δW must be negative. They have also established procedures for determining the ξ which gives the minimum δW. A related analysis has been presented by Hain, Lüst, and Schlüter (9). The particle drifts which actually occur when δW is negative have been analyzed by Rosenbluth and Longmire (16), providing a microscopic explanation of hydromagnetic instabilities.

We shall confine our attention here to the macroscopic picture, considering first the energy change δW_S at an interface, where the plasma pressure in the unperturbed equilibrium state changes discontinuously across some surface, S, parallel to the lines of force. We derive the value of δW_S, computing the change in energy resulting when the surface layer is replaced by a flexible membrane, and the membrane is slowly perturbed, with the plasma on each side of the membrane remaining in equilibrium. Let ξ_n be the perturbation in a direction normal to the surface. There will be a force **F** per unit area across the interface, proportional to ξ_n. The total work done on the fluid in the course of the displacement ξ_n is given by

$$\delta W_S = -\frac{1}{2}\int \xi_n \cdot \mathbf{F}(\xi_n) dS \tag{4-5}$$

integrated over the surface. As we shall see below, the total pressure across the interface in the equilibrium state is $p + B^2/8\pi$. As ξ_n increases, the pressures on the two sides of the surface change in a different way, since $\nabla_n(p + B^2/8\pi)$ is different on the two sides; we denote by ∇_n the component of the gradient normal to S. Evidently $-\mathbf{F}(\xi_n)$ is the product of ξ_n times this increase in gradient as the surface is crossed in the direction of increasing ξ_n. Hence we have

$$\delta W_S = \frac{1}{2}\int \xi_n^2 \left\langle \nabla_n \left(p + \frac{B^2}{8\pi}\right)\right\rangle dS \tag{4-6}$$

where $\langle X \rangle$ denotes the change of some quantity X across the surface, defined as $X(\xi_n) - X(-\xi_n)$ as ξ_n approaches zero.

There will also be a change of potential energy, δW_p, resulting from deformations within the plasma. To simplify the problem we shall assume that the plasma is incompressible. In the simple problems where we shall compute δW_p, the ξ which gives the minimum value of δW is that in which $\nabla \cdot \xi$ vanishes, and even in more general situations this restriction will not affect very greatly the criterion for stability. The computation of δW_p is similar in principle to that of δW_S, but more involved, yielding (2)

$$\delta W_p = \frac{1}{2}\int \left\{ \frac{(\delta B)^2}{4\pi} - \frac{1}{4\pi}\mathbf{j} \cdot \delta \mathbf{B} \times \xi - (\xi \cdot \nabla \phi)(\xi \cdot \nabla \rho) \right\} dV \tag{4-7}$$

where dV is an element of volume; $\delta \mathbf{B}$, obtained by combining equations (2-18) and (2-36), and integrating over dt, is

$$\delta \mathbf{B} = \nabla \times (\xi \times \mathbf{B}) \tag{4-8}$$

In addition to δW_S and δW_p one must also consider δW_v, the change of magnetic energy in any vacuum regions. The value of $\delta \mathbf{B}$ within a vacuum is determined uniquely by Maxwell's equations, together with the value of $\xi \times \mathbf{B}$ at any interface between plasma and vacuum. In those problems where we shall compute δW no vacuum fields are considered.

4.2 Plane System

We consider first the situation where all quantities are functions of x only. This idealized case has the advantage that it can be treated completely with relatively simple mathematics, and the basic concepts of plasma diamagnetism, force-free fields, and flute or interchange instabilities clearly demonstrated.

a. Equilibrium. Except in the trivial case where all components of $\mathbf{B}$ are constant, B_x must vanish. This result follows from equation (4-1), where $B_x\partial B_y/\partial x$ and $B_x\partial B_z/\partial x$ must vanish since ∇p and ∇B^2 have no components in the y or z directions. Hence the $\mathbf{B}\cdot\nabla\mathbf{B}$ term vanishes in equation (4-1) and we can integrate over dx to obtain

$$p + \frac{B^2}{8\pi} = \text{constant} \tag{4-9}$$

With this simple geometry $B^2/8\pi$ may be regarded as a magnetic pressure, and the sum of the material and magnetic pressures is constant. Thus a confined plasma tends to be diamagnetic, with the magnetic pressure reduced below its vacuum value by the presence of the plasma. Evidently p and B can be arbitrary functions of x, provided only that equation (4-9) is satisfied.

Moreover, the direction of $\mathbf{B}$ in the yz plane can be an arbitrary function of x without affecting equation (4-9). If the magnitude, B, of the magnetic field is independent of x, there is no force on the plasma even if the ratio B_y/B_z varies in an arbitrary way. This is an example of a so-called "force-free field." While a vacuum field is also force free, this name is usually reserved for a field in which a current, $\mathbf{j}$, is present, satisfying the requirement

$$\mathbf{j} = \nabla \times \mathbf{B} = \alpha(\mathbf{r})\mathbf{B} \tag{4-10}$$

where $\alpha(\mathbf{r})$ is some function of position, which must be constant along each line of force if $\nabla \cdot \mathbf{j}$ is to vanish. Evidently if the current is parallel to the lines of magnetic force, as specified by equation (4-10), the ponderomotive force $\mathbf{j} \times \mathbf{B}$ vanishes. In the plane case, where all quantities are functions of x only, the most general force-free field (apart from the vacuum field in which $\mathbf{B}$ is constant) is evidently that in which B_x vanishes, B is constant, and the ratio B_y/B_z is an arbitrary function of x. If we let B_y and B_z equal $B \sin \theta$ and $B \cos \theta$, respectively, equation (4-10) yields

$$\frac{d\theta}{dx} = \alpha(x) \tag{4-11}$$

A magnetic field whose direction in the yz plane changes continuously with x is sometimes called a "shear field."

Next we consider the electric field present in this plane case. If we assume that $\mathbf{v}$ vanishes, equation (2-21) determines $\mathbf{E}$ directly in terms of p_i. If we express $\mathbf{E}$ in terms of the potential, U, and let p_i equal n_ikT, taking T to be constant, we may integrate over x to find

$$\log n_i + \frac{Ze}{ckT} U = \text{constant} \tag{4-12}$$

Thus if $\mathbf{v}$ equals zero, the electric potential must be such that the density of positive ions obeys the Boltzmann law. Evidently in this case the ponderomotive force produced by a gradient of $\mathbf{B}$ must be exerted entirely on the electrons, which carry virtually all the current.

This result may be compared with the electric field found in an isothermal atmosphere without a magnetic field but subject to a gravitational field (or to a uniform acceleration). We must now retain the gravitational terms in equations (2-20) and (2-21). Since $p_i = p_e/Z = p/(1 + Z)$, if the electron and positive ion temperatures are equal, equation (2-21) yields, if we eliminate p_i by means of equation (2-20),

$$-\frac{Ze}{c}\nabla U = \frac{Z}{1+Z}m_i\nabla\phi \tag{4-13}$$

Thus the vertical electric force per ion, given by the left-hand side of this equation, cancels a fraction $Z/(1+Z)$ of the gravitational force in a positive ion, and provides a net downward force on the positive ions just equal to that on the electrons. Evidently this electric field is required to preserve electric neutrality.

b. Stability, Isotropic Pressure. If the magnetic field is uniform and the velocity distribution is Maxwellian at a uniform temperature, it is clear that no instabilities can be present; there is no state of lower energy or greater entropy to which the system can go. If the plasma is confined, and p varies with x, the situation alters. Evidently collisions produce diffusion across the lines of force, and in the absence of collisions it is conceivable that more complex effects in a plasma may somehow produce the same effect, through the action of instabilities, turbulence, etc. Whether or not a confined plasma is intrinsically unstable is still an unsolved problem.

On the basis of the macroscopic equations given in Chapter 2, based on an isotropic pressure, no instabilities arise in a plasma in plane equilibrium, provided that the gravitational force vanishes; under these conditions the integral in equation (4-7) becomes a sum of squares if ξ vanishes at the limits of integration. Physically it is clear that compressing one part of the magnetic field and expanding another increases the energy, since the rising pressure of the compressed portion requires more energy than is liberated from the expansion, at reduced pressure, of the expanded portion.

In the special case that the lines of force are all parallel in the equilibrium state, perturbations are possible in which δW vanishes. We have already taken $\nabla\cdot\xi$ equal to zero. If we assume also that $\mathbf{B}$ is everywhere parallel to the z axis, and that $\partial\xi/\partial z$ is zero, then there will be no bending of the lines of force in the perturbation and no change in magnetic energy. In the xy plane the motion is that of an incompressible fluid, and is independent of z. The lines of force move as rigid rods, and through any fluid element the magnetic intensity, $\mathbf{B}$, is constant both in magnitude and direction. Evidently the plasma is neutral against such perturbations, which can exchange regions where $\mathbf{B}$ is strong for regions where $\mathbf{B}$ is weak. Such perturbations, in which some lines of force move in one direction, while others move oppositely, are called "interchanges."

If electric currents parallel to $\mathbf{B}$ are present everywhere, and, as a result, $\mathbf{B}$ rotates in the yz plane as x increases, the plasma is no longer neutral against such interchanges. The direction of the lines of force in the yz plane is now different for different x, and, apart from trivial motions of uniform translation, the lines of force cannot interchange their positions, moving as rigid rods, because other lines of force are in the way. Evidently a shear magnetic field tends to have greater stability than one without shear.

When the direction of $\mathbf{B}$ is everywhere the same, an instability can arise if the plasma is supported by a magnetic field against a gravitational force $\rho\mathbf{g}$ per cm^3, or if the magnetic field is accelerating the plasma against the equivalent reaction force, $-\rho d\mathbf{v}/dt$. For the case where a sharp interface separates two regions, with differing densities and field strengths, we may compute δW by use of equation (4-6) above. If we retain the gravitational term in equation (2-20), equation (4-1) becomes in this one-dimensional case

$$\frac{d}{dx}\left(p + \frac{B^2}{8\pi}\right) = -\rho g \qquad (4\text{-}14)$$

The gravitational acceleration $\mathbf{g}$ is assumed to be directed toward decreasing x. Equation (4-6) now yields

$$\delta W_S = -\frac{1}{2}\langle\rho\rangle g \int \xi\delta^2\, dS \qquad (4\text{-}15)$$

and is negative if the density of the upper layer exceeds that of the lower layer. If positive x is taken to be the direction of **g**, the minus signs in equations (4-14) and (4-15) are changed to plus signs; however, the definition of $\langle\Delta\rho\rangle$ following equation (4-6) is also modified, with an additional change of sign introduced, and the final sign of δW_S remains unchanged. Changes of energy in the volume layers above and below the interface must also be considered. If we choose a $\boldsymbol{\xi}$ which is constant along the lines of force, these lines will move as rigid rods, and there will be no change of magnetic energy. Volume changes of material energy arising from the last term in equation (4-7) will also be small, provided that $2\pi/\kappa$, the wavelength of the perturbation, is very small compared to the scale height $\rho/\nabla\rho$. Hence we see that δW is negative if a dense plasma is supported against gravity by a lighter plasma, provided that the direction of **B** is everywhere uniform; this situation is therefore unstable. This same conclusion follows if a lighter plasma accelerates a denser plasma by pushing against it.

The unstable perturbation leads to a corrugation or fluting of the interface, with the flutes parallel to the lines of force. Hence the instability of an interface against these short-wavelength perturbations is frequently called a "flute" instability. Since this perturbation leads to an interchange of lines of force, with no bending of the lines, such an instability is sometimes also referred to as an "interchange" instability.

The rate at which the disturbance grows may be computed from equation (4-4). The simplest $\boldsymbol{\xi}$ which does not alter the magnetic energy but which goes to zero away from the interface is

$$\begin{aligned} \xi_x &= A\, e^{\pm\kappa x} \sin \kappa y \\ \xi_y &= \pm A\, e^{\pm\kappa x} \cos \kappa y \\ \xi_z &= 0 \end{aligned} \tag{4-16}$$

with the minus sign taken above the interface, the plus sign below. With this perturbation we obtain

$$\omega^2 = -\frac{g\kappa\langle\rho\rangle}{2\bar{\rho}}, \tag{4-17}$$

where $\bar{\rho}$ denotes the average of ρ on the two sides of the interface, while $\langle\rho\rangle$ denotes the difference of ρ on the two sides. Since the perturbation $\boldsymbol{\xi}$ assumed in equation (4-16) is the solution given by the normal mode approach, equation (4-17) is the usual result for Rayleigh-Taylor instability. Evidently the disturbances of shortest wavelength grow most rapidly.

If **B** changes direction across the interface or in the adjacent volume, as a result of electrical currents along the lines of force, these results will alter. In this case it is not possible to interchange lines of force in the upper and lower layers; any perturbation which interchanges fluid at different values of x must bend the lines of force and increase the magnetic energy. The magnitude of this effect is simply computed if we assume that **B** makes an angle θ with the z axis above the interface, and an angle $-\theta$ below. If $\boldsymbol{\xi}$ is given by equation (4-16), the increase in magnetic energy may be obtained by substituting equation (4-8) into (4-7). It is readily shown that the resulting ratio of δW_p to δW_S is $\kappa B^2 \sin^2 \theta/(-\ 2\pi g\langle\rho\rangle)$. Hence δW will be positive for small wavelengths (large κ), but for sufficiently large wavelengths instability results as before. This is because for very long wavelengths the perturbation involves relatively little bending of the lines of force. Stabilization of the shorter wavelengths is a rather general property of shear fields.

c. Stability, Anisotropic Pressure. When $p_\perp$ and $p_\parallel$ are different, even a uniform medium may be unstable. Since the unstable perturbations produce gradients along **B**, the macroscopic equations are not strictly applicable, and an accurate analysis must deal directly with the velocity distribution function. In particular, magnetic mirrors will develop during the course of the perturbation, or may be enhanced if present initially. The presence of trapped particles in these mirrors cannot be taken into account by macroscopic equations, based on simplifying assumptions about the pressure tensor.

However, analysis of this situation by means of the macroscopic equations does give some insight into the situation. We consider, therefore, one case where these equations do give the correct result—the instability of an initially uniform plasma in which $p_{\parallel}$ exceeds $p_{\perp}$ by more than $B^2/4\pi$. If the gravitational potential vanishes, and if **B** and ρ are uniform in the equilibrium state, then we see from equation (4-2) that the change of potential energy is given by

$$\delta W = \frac{1}{8\pi}\int (\delta B)^2 d\tau + \frac{1}{2}\int \delta p_{\parallel} d\tau + \int \delta p_{\perp} d\tau \tag{4-18}$$

The $\mathbf{j}\cdot\delta\mathbf{B} \times \boldsymbol{\xi}$ term which appeared in equation (4-7) is absent here since **j** vanishes in the equilibrium state.

We let **B** be parallel to the z axis, and consider a perturbation in which $\boldsymbol{\xi}$ is parallel to the x axis. Evidently $\nabla\cdot\boldsymbol{\xi}$ vanishes if $\partial\xi_x/\partial x$ is zero. If we take ξ_x to be a function of z, independent of y as well as x, the perturbation is of the type occurring in an Alfvén wave. It is evident physically, and may be verified from equation (4-8), that $\delta B_y = \delta B_z = 0$, and only δB_x need be considered. Since the motion is incompressible, $\delta p_{\parallel}$ and $\delta p_{\perp}$ may be replaced by $nk\delta T_{\parallel}$ and $nk\delta T_{\perp}$, respectively, with $T_{\parallel}$ and $T_{\perp}$ determined from equations (1-32) and (1-35), respectively. The value of $(\delta B)^n/B^n$ that occurs in the expression for $(\delta T_{\parallel})/T_{\parallel}$ and $(\delta T_{\perp})/T_{\perp}$ may be found from the simple result

$$\frac{\delta B^n}{B^n} = \{B^2 + (\delta B_x)^2\}^{n/2} - B_z^{\,n} \approx \frac{n}{2}\frac{(\delta B^2)}{B^2} \tag{4-19}$$

since the initial **B** is entirely in the z direction. Combining equations (4-18) and (4-19) yields

$$\delta W = \frac{1}{2B^2}\int (\delta B)^2 d\tau \left\{\frac{B^2}{4\pi} - nkT_{\parallel} + nkT_{\perp}\right\} \tag{4-20}$$

Evidently the perturbation will be unstable if

$$p_{\parallel} - p_{\perp} > \frac{B^2}{4\pi} \tag{4-21}$$

This instability, pointed out originally by Parker (14), is sometimes called the "firehose" instability. If $p_{\parallel}$ is much larger than either $p_{\perp}$ or $B^2/4\pi$, bending the lines of force, and thus increasing their length, will decrease $p_{\parallel}$ more than it will increase either the magnetic energy or the transverse kinetic energy. From the standpoint of the forces involved, the centrifugal force associated with a curving path tends to accentuate the curves, when $p_{\parallel}$ is predominant.

Since the perturbed magnetic field varies in magnitude along the lines of force, some trapping of particles occurs as the instability develops. However, the analyses by Chandrasekhar, Kaufman, and Watson (4) and by Sagdeyev and his colleagues (17) indicate that a more exact treatment still yields inequality (4-21) as a criterion for instability. This is not too surprising, since the change of B is of second order in ξ^2, and the trapping does not affect those particles which provide the driving force for the instability; i.e., the particles which are moving predominantly along the lines of force.

There is an alternate mode of instability, called the "mirror" instability, that can develop when $p_{\perp}$ much exceeds $p_{\parallel}$ in an initially uniform medium. In this mode $\partial\xi_x/\partial x$ is finite; from equation (4-8) we see that δB_z equals $-B_z\partial\xi_x/\partial x$, and the strength of the total magnetic field varies linearly with ξ, instead of quadratically as in the firehose instability. If we assume also that $\partial\xi_x/\partial x$ oscillates slowly with increasing z, then the perturbation leads to the development of magnetic mirrors—with a single line of force passing alternately through regions of diminished field and regions of enhanced field. Particles whose velocities are predominantly transverse to **B** will be trapped in these mirrors, and provide a dominant driving force for instability, since the kinetic energies of these particles decrease as the instability develops. Chandrasekhar (4), Sagdeyev (17), and their colleagues have shown that the criterion for instability, not obtainable from the simple macroscopic equations, is

$$p_{\perp} - p_{\parallel} > \frac{p_{\parallel}}{p_{\perp}} \frac{B^2}{8\pi} \tag{4-22}$$

Rough observational confirmation of the mirror instability has been obtained by Post and Perkins (15), who observed a very great increase in the diffusion rate from a magnetic mirror under conditions generally consistent with inequality (4-22).

4.3 Cylindrical System

When all quantities are functions only of r, the distance from the cylindrical axis, the situation differs from the plane case in that the lines of force may be curved in the equilibrium situation. As a result the equilibrium conditions are somewhat modified, and important new types of instability appear.

a. Equilibrium. To apply equation (4-1) in this geometry we express equation (2-19) for $\mathbf{j}$ in cylindrical coordinates, with $\partial \mathbf{E}/\partial t$ again set equal to zero. We have the familiar results

$$4\pi j_r = \frac{1}{r}\frac{\partial B_z}{\partial \theta} - \frac{\partial B_\theta}{\partial z} \tag{4-23}$$

$$4\pi j_\theta = \frac{\partial B_r}{\partial z} - \frac{\partial B_z}{\partial r} \tag{4-24}$$

$$4\pi j_z = \frac{1}{r}\frac{\partial}{\partial r}(rB_\theta) - \frac{1}{r}\frac{\partial B_r}{\partial \theta} \tag{4-25}$$

If we assume that $\mathbf{B}$ is a function of r only, then from equation (4-23) it follows that j_r must vanish. If we also stipulate that $\partial p/\partial z$ and $\partial p/\partial \theta$ vanish, then from equation (4-1) we see that B_r must also vanish except in the trivial case where both j_θ and j_z vanish. If we set B_r equal to zero in equations (4-24) and (4-25), the r component of equation (4-1) becomes

$$\frac{\partial}{\partial r}\left\{p + \frac{B_\theta^2 + B_z^2}{8\pi}\right\} + \frac{1}{4\pi}\frac{B_\theta^2}{r} = 0 \tag{4-26}$$

If B_θ vanishes, the sum of the material and magnetic pressures must be constant, as in equation (4-9). If B_θ differs from zero, however, the stresses due to tensions along the lines of force affect the equilibrium.

In the general case equation (4-26) admits a simple integral, provided that $p(r)$ vanishes for r equal to or greater than some limiting radius, R. To derive this integral we transform equation (4-26) by the substitution of S and I for r and B_θ, where

$$dS = 2\pi r dr \tag{4-27}$$

$$B_\theta = \frac{2I}{r} \tag{4-28}$$

Evidently S is the cross-sectional area of the plasma interior to r, and I is the current in the z direction through this cross section. Equation (4-26) then becomes

$$S\frac{d}{dS}\left\{p + \frac{B_z^2}{8\pi}\right\} = -I\frac{dI}{dS} \tag{4-29}$$

Evidently dI/dS equals j_z, the current density in the z direction. On integrating by parts, we find

$$\frac{1}{2}I^2(R) = NkT + \frac{1}{8\pi}\int_0^R (B_z^2(r) - B_z^2(R))2\pi r dr \tag{4-30}$$

where we have assumed that T is constant and that, as a result

$$\int_0^R pdS = kT\int_0^R ndS = NkT \tag{4-31}$$

The quantity N is the total number of particles per centimeter length of the confined plasma. The special case of equation (4-30) when B_z vanishes was first derived by Bennett (1). The derivation given here is largely due to King (10). The confinement of a current filament by its own self-magnetic field is known as the "pinch effect." Here we shall refer to a plasma confined by its own B_θ field as a "self-pinched" plasma, and denote a plasma confined by an external B_z field as "externally-pinched." The self-pinched discharge is sometimes called a "longitudinal pinch," since the plasma current is longitudinal, while the externally-pinched discharge is called an "azimuthal pinch" or "theta pinch."

As in the plane plasma a force-free field with cylindrical symmetry is possible, in which ∇p is zero everywhere, and where the direction of **B** rotates as r increases. In such a field B_z^2 and I^2 must be related by equation (4-29), with p set equal to zero, but are otherwise arbitrary, except that I must vanish as S goes to zero. While in the plane case B is constant in a force-free field, with only the direction of **B** changing, in this cylindrical case the magnitude of **B** will change with increasing r. If we impose the condition that B_z and j_z both vanish together for r equal to some R, and also for all greater values of r, then for r greater than R, I is independent of S and B_θ varies as $1/r$. In the simplest such system j_z is constant for r less than R, and B_z^2 decreases linearly with S in this range of r. While such a system is force-free, the stresses are not zero. The lines of force near the axis, where **B** is predominantly in the z direction, tend to repel each other. The counter stress needed is provided by the B_θ field at greater distances, where the tension along the lines of force tends to pull these lines inward.

The radial electric field E_r and the velocity, **v**, in a self-pinched plasma still remain to be determined. These are related by equation (2-21), which yields, with neglect of the η**j** term,

$$E_r - v_z B_\theta + v_\theta B_z - \frac{c}{en_e}\frac{dp_i}{dr} = 0 \qquad (4\text{-}32)$$

Steady-state conditions do not suffice to determine these quantities and we must inquire into the origin of the plasma. The velocity obtained from equation (4-32) is perpendicular both to **B** and to **r**. It is clear from the equation of motion, (2-11), that a change in this velocity can be produced only by a radial current, j_r, which can produce a pondermotive force in the direction $\mathbf{r} \times \mathbf{B}$. Since any appreciable radial current will produce a large electrostatic field, E_r, one would expect **v** to remain small. We shall show that this is actually the case. The analysis will be simplified by the assumption that B_z vanishes; the same result follows also in the more general case.

The z component of equation (2-11) is

$$\rho \frac{\partial v_z}{\partial t} = j_r B_\theta \tag{4-33}$$

In addition the component of equation (2-19) in the r direction yields

$$4\pi j_r + \frac{\partial E_r}{c^2 \partial t} = 0 \tag{4-34}$$

Equation (4-34) may be used to eliminate j_r from equation (4-33).

To permit integration of the resultant equation over time we shall assume that ρ/B_θ is constant with time during the formation of the pinched discharge. This assumption is not very realistic for a self-pinched discharge, but should give an approximate indication of what may be expected.

With this assumption, equation (4-33) yields

$$E_r = -\frac{4\pi\rho c^2}{B_\theta} v_z \tag{4-35}$$

provided we assume that initially both E_r and v_z are zero. If now we combine equations (4-35) and (4-32), (with $\partial p_i/\partial r$ replacing dp_i/dr) and eliminate E_r, we obtain

$$v_z = -\frac{c}{Ken_eB_\theta} \frac{\partial p_i}{\partial r} \tag{4-36}$$

where K is the dielectric constant, given in equation (2-33). If E_r were zero, v_z would be given by equation (4-36) with K equal to one. It is evident that for large K the radial electrostatic field almost exactly cancels the velocity which the position-ion pressure gradient would otherwise produce.

A similar conclusion has been established by Spitzer (20) in a somewhat different case. If a plasma is confined by a magnetic field, and the temperature is increased, it might be supposed that the material velocity $\mathbf{v}$ transverse to $\mathbf{B}$ and to ∇p_i might increase as ∇p_i is increased, in accord with equation

(2-23). However, in this case also the acceleration must be produced by electric currents along the pressure gradient, and these currents produce an electrostatic field parallel to ∇p_i, the residual velocity amounting, as in equation (4-36), to only $1/K$ times its value in the absence of the electrostatic field.

b. Stability. When the lines of force are curved in the equilibrium situation, an instability can be present even in the absence of acceleration. As in the previous section, we consider here the instability of an interface, where the pressure changes discontinuously. Since the pressure of a B_z field tends to stabilize the gas, we shall first consider the self-pinched cylindrical plasma, with B_z equal to zero.

The change of potential energy resulting from perturbations in an interface is given in equation (4-6). As in the plane case, $p + B^2/8\pi$ is continuous across the interface. We assume that the interface is characterized by some value of $\langle p \rangle$; for a confined plasma p decreases outwards and $\langle p \rangle$ will be negative. The value of $\partial B^2/\partial r$, obtained from equation (4-26), is $-2B_\theta^2/r$, since $\partial p/\partial r$ is assumed zero. Hence we have

$$\left\langle \nabla_n \left(p + \frac{B^2}{8\pi} \right) \right\rangle = -\frac{1}{r} \left\langle \frac{B_\theta^2}{4\pi} \right\rangle \tag{4-37}$$

Using again the fact that $p + B_\theta^2/8\pi$ is continuous, we obtain from equation (4-6)

$$\partial W_S = \frac{\langle p \rangle}{r} \int \xi_n^2 \, dS \tag{4-38}$$

We see that δW_S is negative if $\langle p \rangle$ is negative; i.e., if p just outside the cylindrical boundary is less than the value just inside.

Instability will result for any ξ which makes the volume contribution to δW negligible. Hence we again consider a ξ for which $\nabla \cdot \xi = 0$, and $\mathbf{B} \cdot \nabla \xi$ vanishes. If the wavelength of the perturbation is sufficiently small compared to r, these requirements are satisfied approximately by equation (4-16), with r, z, and θ replacing x, y, and z, respectively. Change of potential energy in the volume of the gas and in the vacuum outside are both negligible, and in place of equation (4-17) we now obtain

$$\omega^2 = \frac{\kappa \langle p \rangle}{r\bar{\rho}} \tag{4-39}$$

valid if κr is much greater than unity. This is another example of the flute instability, which in theory is always present at an interface if the lines of force are concave towards the region of greater plasma pressure and if the magnetic field has no shear.

If κr is small δW_p and δW_v are no longer negligible. Since the equilibrium state is independent of θ and z, we may conveniently take Fourier components of $\boldsymbol{\xi}$ in the θ and z direction, writing

$$\boldsymbol{\xi} = \mathbf{f}(\mathbf{r})e^{im\theta+i\kappa z-i\omega t} \tag{4-40}$$

Evidently $\boldsymbol{\xi}$ is a periodic function of θ, and hence m must be an integer. The criteria for stability differ for each m, and must be determined by a detailed analysis. A number of such analyses have been carried out, including normal-mode analyses by Kruskal and Schwarzschild (12), Kruskal and Tuck (13), Shafranov (18), and Tayler (22). Some of the salient results of these analyses are given here for reference, retaining the restriction to a uniform cylindrical plasma, with a vacuum field outside.

For $m = 0$ the plasma is unstable for every κ. For κr small the distorted plasma column resembles somewhat a string of sausages, and hence instability in this mode is sometimes called the "sausage" instability. If the plasma pressure is uniform in the unperturbed state, with a vacuum outside a cylinder of radius r, the growth rate of the $m = 0$ mode, in the limit of small κr, is given by

$$\omega^2 = -\kappa^2 \frac{\gamma p}{(\gamma - 1)\rho} \tag{4-41}$$

where now p and ρ refer to quantities in the plasma. For the disturbance with $m = 1$, the perturbed plasma develops a wavy, or kinked, appearance, and hence this mode of instability is sometimes called the "kink" instability. In the limit of small κr the growth rate for this mode is given by

$$\omega^2 = -\kappa^2 \frac{2p}{\rho}\left(1 - \gamma + \ln \frac{2}{\kappa r}\right) \tag{4-42}$$

where γ is Euler's constant, equal to 0.5772. The modes for $m = 2$ or higher are stable for low κr, but, in common with the modes for $m = 0$ and 1, obey equation (4-39), with $\bar{\rho}$ replaced by $\rho/2$, for κr appreciably greater than m^2.

As in the plane situation, these instabilities tend to be stabilized by shear fields. With B_θ outside the plasma the greatest shear is achieved if an axial magnetic field, B_z, is assumed present within the plasma, but not outside. This stabilization is most complete for the $m = 0$ mode. The criterion for stabilization may be obtained by considering the forces involved. The destabilizing force is produced by the B_θ field just outside the plasma; this field varies as $1/r$ for the radial perturbation in the $m = 0$ mode, and any compression of the plasma, initially in equilibrium, will produce an increase in the inwards force and will tend to grow. An expansion is similarly unstable. The gas pressure will not oppose the instability, if the motion is incompressible, with the compression at some value of z being offset by an expansion at other axial positions. The B_z field inside the plasma, on the other hand, tends to resist the instability. Evidently B_z varies as $1/r^2$, since the axial flux remains constant. Stability will result if at the plasma boundary B_z^2 increases faster than B_θ^2 as the plasma contracts. Since the relative change of B_z^2 is twice that of B_θ^2 for a given change of r, it follows that an axial magnetic field inside the plasma will stabilize the sausage instability if, at the plasma boundary

$$B_z^2 \geq \frac{1}{2} B_\theta^2 \tag{4-43}$$

The kink instability is more difficult to stabilize, since the B_z field lines are somewhat stretched by the perturbation but are not much compressed; in a plane of constant z, the perturbation for $m = 1$ gives a uniform translation with no distortion. As in the plane situation, stability is achieved only for sufficiently large κ. More specifically, if B_z vanishes outside the plasma cylinder, with B_θ zero inside, instability will still be present if

$$\kappa r < 2e^{1-\gamma-B_z^2/B_\theta^2} \tag{4-44}$$

where B_z is evaluated just inside the plasma boundary and B_θ, just outside; γ is again Euler's constant. Even in the extreme case where B_z inside the plasma is about equal to B_θ outside, and the plasma pressure is vanishingly small, instability sets in for all κr less than about 1.12. In the situation where B_z has the same value outside and inside the plasma, with B_θ at the plasma boundary much less than B_z, the condition for instability becomes, to first order in B_θ/B_z

$$\kappa r < \frac{B_\theta}{B_z} \tag{4-45}$$

Stabilization of the kink instability, for a plasma cylinder with uniform pressure, can, in theory, be achieved if a perfectly conducting cylinder surrounds the plasma. However, if the region of pressure gradient is analyzed in more detail, interchange instabilities within this layer tend to be present; these instabilities correspond to perturbations of high m. To stabilize the self-pinched discharge against hydromagnetic instabilities within the entire volume of the plasma requires rather special conditions; for example, the plasma will be stable if B_θ decreases outwards more rapidly than $1/r$, a configuration that may be achieved if a hollow cylindrical plasma surrounds an axial conductor which carries electric current—sometimes called the "hard-core pinch" (5).

Some experimental confirmation of these instabilities has been obtained in fully ionized gases. Self-pinched discharges have been observed to be unstable for the modes $m = 0$ and $m = 1$, although in the pinches studied by Curzon et al. (7) the Rayleigh-Taylor instability arising during compression of the pinch is apparently about as important as the instability due to curvature of the lines of force. In plasmas with a uniform B_z present as well as B_θ, the $m = 1$ instability has been observed by Kruskal et al. (11) and by Dolgov-Saveliev et al. (8) to occur under conditions predicted by equation (4-45). However, the interchange instabilities of high m, predicted by the theory for situations in which p changes continuously across the plasma, have not been observed.

4.4 Axisymmetric System

With more complicated systems the number of possibilities becomes larger and larger, and hence of less general interest. Specific systems can always be analyzed by direct numerical methods. In the present subsection we consider only a few general results on the equilibrium of axisymmetric configurations, primarily those in which either the magnetic field lines or the lines of electric current flow are circles about the axis of symmetry. Particle confinement in axisymmetric systems has already been demonstrated in Section 1.2, provided that **B** has an appreciable component parallel to the axis of symmetry.

First we treat systems in which the magnetic field lines are circles; i.e., B_r and B_z vanish and B_θ is independent of r. This is the one case where the confinement discussion in Section 1.2 does not apply, since $\Phi(r, z)$ vanishes. The macroscopic equation (4-1) now yields a particularly simple result. If we take the curl of both sides of equation (4-1) the left-hand side vanishes; the θ component of the right-hand side yields

$$\frac{\partial}{\partial z}(j_z B_\theta) + \frac{\partial}{\partial r}(j_r B_\theta) = 0 \qquad (4\text{-}46)$$

If we substitute equations (4-23) and (4-25) for j_r and j_z we obtain

$$\frac{1}{r}\frac{\partial B_\theta^2}{\partial z} = 0 \tag{4-47}$$

If r is not infinite, B_θ must be independent of z, and hence p also must be independent of z. Thus a plasma cannot be confined in a finite volume by a simple toroidal B_θ field, and the externally pinched plasma is not in equilibrium if bent into a toroidal form. Physically this result may be attributed to particle drifts associated with the inhomogeneity of the B_θ field. As a result of these drifts, electric currents appear in the z direction, and if B_θ varies with z, these currents will possess a divergence. This same result may also be shown directly from the macroscopic equations; if equation (2-20) is solved for $\mathbf{j}$, with a pure toroidal B_θ field and the gravitational potential ignored, $\nabla \cdot \mathbf{j}$ does not vanish unless equation (4-47) is satisfied.

Next we consider the converse case, where the lines of current flow are circles about the axis of symmetry. Now B_θ vanishes identically under these conditions. If the current, j_θ, is confined to the gas alone, it is readily shown that equilibrium is not possible; to hold the configuration together currents in external conductors are required. The plasma configuration depends both on the geometry and on the relative magnitude of these external currents. A very wide variety of equilibrium situations are possible. One limiting case is that in which the external currents have their minimum value for equilibrium; this situation corresponds to the self-pinched plasma when it is bent into toroidal form.

Another limiting case of axisymmetric plasmas with circular currents is a plasma confined between two magnetic mirrors in which the external currents much exceed the plasma current. As we have seen in Section 1.3, confinement between magnetic mirrors is possible only if the velocity distribution is anisotropic. In this more general case ∇p in equation (4-1) must be replaced by $\nabla \cdot \mathbf{\Psi}$, where $\mathbf{\Psi}$ now has the components $p_\perp$ in the two directions perpendicular to $\mathbf{B}$, and $p_\parallel$

parallel to **B**. Since the directions in which $\boldsymbol{\Psi}$ is diagonal rotate with position in this geometry, $\nabla \cdot \boldsymbol{\Psi}$ may be finite even if $p_{\parallel}$ and $p_{\perp}$ are constant. After some algebra we find that the component of this generalized equation (4-1) which is parallel to **B** yields

$$\nabla_{\parallel} p_{\parallel} = (p_{\parallel} - p_{\perp}) \frac{\nabla_{\parallel} B}{B} \tag{4-48}$$

where $\nabla_{\parallel}$ denotes the component of the gradient in the direction of the magnetic field. Since $p_{\perp}$ equals $nmw_{\perp}^2/2$, equation (4-48) reduces, in the limit where $p_{\parallel}$ is much less than $p_{\perp}$, to the familiar equation of hydrostatic equilibrium (equation (4-14) with B set equal to zero), provided we replace the gravitational acceleration g by the acceleration of a diamagnetic particle in a magnetic gradient—see equation (1-24).

Equilibrium configurations in which the electric currents are circles about the axis of symmetry and which are intermediate between a mirror and a self-pinched toroidal plasma are also possible (6). In these an external B_z field contributes substantially to the confinement, but the current in the toroidal plasma is sufficiently great to weaken or even to reverse the B_z field threading the plasma.

More general axisymmetric fields are also possible, of course, in which neither the current lines nor the lines of magnetic force are circles. If a small j_θ is assumed to pass through an externally pinched toroidal plasma, equilibrium becomes possible (3, 19); the lines of force are no longer closed and current divergences can be neutralized by currents parallel to **B**. The same effect can be produced without plasma currents by more complicated external fields, without axial symmetry (21). These complex fields, which are utilized in the stellarator, are of importance for stable plasma confinement but are beyond the scope of this book.

References

1. Bennett, W., *Phys. Rev.*, **45**, 890 (1934).
2. Bernstein, I. B., E. A. Frieman, M. D. Kruskal, and R. Kulsrud, *Proc. Roy. Soc.* (*London*) **A244**, 17 (1958).

3. Biermann, L., K. Hain, K. Jörgens, and R. Lüst, *Z. Naturforsch.*, **12a**, 826 (1957).
4. Chandrasekhar, S., A. N. Kaufman, and K. M. Watson, *Proc. Roy. Soc. (London)* **A245**, 435 (1958).
5. Colgate, S. A. and H. P. Furth, *Phys. Fluids*, **3**, 982 (1960).
6. Christofilos, N. C.; *Proc. Second United Nations Internat. Conf. on Peaceful Uses of Atomic Energy* (United Nations, Geneva, 1958) Vol. **32**, p. 279.
7. Curzon, F. L., A. Folkierski, R. Latham, and J. A. Nation, *Proc. Roy. Soc. (London)* **A257**, 386 (1960).
8. Dolgov-Saveliev, G. G., V. S. Mukhovatov, V. S. Strelkov, M. N. Shepelov, and N. A. Yavlinski, *J. Exptl. Theoret. Phys. (U.S.S.R.)*, **38**, 394 (1960); also published in *Proceedings of the Fourth International Conference on Ionization Phenomena in Gases, Uppsala* (North-Holland Publ. Co., Amsterdam, 1960) Vol. **II**, p. 947.
9. Hain, K., R. Lüst, and A. Schlüter, *Z. Naturforsch.*, **12a**, 833 (1957).
10. King, J. I. F., informal communication (1958).
11. Kruskal, M. D., J. L. Johnson, M. B. Gottlieb, and L. M. Goldman, *Phys. Fluids*, 1, 421 (1958).
12. Kruskal, M. D., and M. Schwarzschild, *Proc. Roy. Soc. (London)* **A223**, 348 (1954).
13. Kruskal, M. D. and J. L. Tuck, *Proc. Roy. Soc. (London)* **A245, 222** (1958).
14. Parker, E. N., *Phys. Rev.*, 109, 1874 (1958).
15. Post, R. F. and W. A. Perkins, *Phys. Rev. Letters*, **6**, 85 (1961).
16. Rosenbluth, M. N. and C. L. Longmire, *Ann. Phys.*, **1**, 120 (1957).
17. Sagdeyev, R. S., B. B. Kadomtsev, L. I. Rudakov, and A. A. Vedyanov, *Proc. Second United Nations Internat. Conf. on Peaceful Uses of Atomic Energy* (United Nations, Geneva, 1958), Vol. **31**, p. 151.
18. Shafranov, V., *Plasma Physics and the Problem of Controlled Thermonuclear Reactions*, edited by M. A. Leontovich, 1958 (English translation published by Pergamon Press, 1959, Vol. **II**, p. 197).
19. Shafranov, V., *J. Exptl. Theoret. Phys. (U.S.S.R.)*, **33**, 710 (1957); *Soviet Phys. JETP*, **6**, 545 (1958).
20. Spitzer, L., *Astrophys. J.*, 116, 299 (1952).
21. Spitzer, L., *Phys. Fluids*, **1**, 253 (1958).
22. Tayler, R. J., *Proc. Phys. Soc. (London)* **B70**, 31 (1957).

CHAPTER 5

Encounters between Charged Particles

If a gas is in thermodynamic equilibrium, collisions between the particles are of little interest, as they do not affect the state of the gas. To analyze non-equilibrium phenomena, however, a quantitative study of collisions is necessary. Two types of non-equilibrium phenomena may be distinguished. A gas may be far from equilibrium, and one may wish to know the rate at which equilibrium is approached. For example, a beam of particles passing through a plasma will be scattered and slowed down; or electrons and positive ions may have different kinetic temperatures, which are gradually approaching each other. A second type is that in which a steady non-equilibrium state is present. The flow of an electric current or the transport of heat across the gas are examples; these phenomena may be analyzed in terms of the "transport coefficients," such as the resistivity, η, and the coefficient of thermal conductivity, $\mathcal{K}$.

In the present chapter the effect of collisions is analyzed and the results are applied to these two types of non-equilibrium phenomena. Electrostatic forces between particles have a much longer range than the forces between neutral atoms. As a result we must consider not so much the close collisions between charged particles, which may completely change the particle velocities, but rather the more distant encounters, each of which produces so small an effect that the term "collision" is scarcely applicable. Such distant encounters are analyzed first, and methods of description presented which are then applied in the subsequent sections of this chapter.

5.1 Distant Encounters

When two charged particles pass by each other, each particle moves in a hyperbola relative to the center of mass of the two particles. In a coordinate system in which the center of mass is stationary, the two particles move in a single plane, called the "orbital plane." The paths of the two particles in this plane are shown in Figure 5.1 for the case of two identical particles; the center of gravity is at point A, while the dashed lines represent the asymptotes of the two hyperbolas.

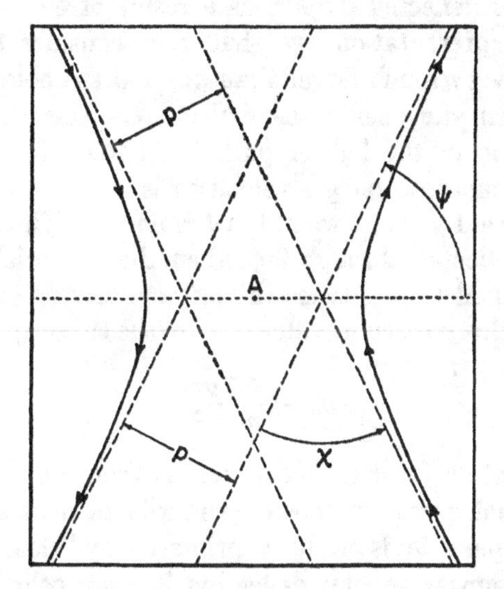

Figure 5.1. Orbits of two identical particles in an encounter. The reference frame is chosen so that the center of gravity, Point A, is stationary.

If the relative velocity, $\mathbf{w}_1 - \mathbf{w}_2$, of the two particles is denoted by $\mathbf{u}$, and the distance of closest approach in the absence of forces is denoted by p, called the "impact parameter," then the angular deflection, χ, of each particle is given by

$$\chi = \pi - 2\psi \qquad (5\text{-}1)$$

where

$$\tan \psi = \frac{Mpu^2}{Z_1 Z_2 e^2} \tag{5-2}$$

Z_1e/c and Z_2e/c are the charges of the two particles, in e.m.u., and M is the reduced mass, defined by

$$\frac{1}{M} = \frac{1}{m_1} + \frac{1}{m_2} \tag{5-3}$$

We are interested in the conditions under which the particle of mass m_1 is deflected by 90°, as a result of encounters. To simplify the presentation, we shall first consider the case in which m_2 is very much larger than m_1, and the heavier particle may be taken stationary; the deflection χ then becomes the true deflection of the lighter particle in the reference system in which the macroscopic gas velocity $\mathbf{v}$ is zero. From equation (5-1) we see that χ is $\pi/2$ when $\tan \psi$ is unity. Thus the lighter particle is deflected through 90° when the potential energy at a distance equal to p is twice the original kinetic energy. We may denote this particular value of p by the subscript 0. Clearly

$$p_0 = \frac{Z_1 Z_2 e^2}{m_1 w_1^2} \tag{5-4}$$

The cross section for such encounters is then πp_0^2.

In normal gases, composed primarily of neutral particles, deflection of particles is produced primarily by "close collisions," with a substantial angular deflection in each collision. If we define as a "close collision" an encounter producing a deflection of 90° or more, the time interval between such close collisions is then a good approximation to the collision time in the gas, which we here denote by t_c. We have

$$t_c = \frac{1}{\pi n_2 w_1 p_0^2} \tag{5-5}$$

where n_2 is the number of particles of mass m_2 per cubic centimeter.

In a gas composed of charged particles equation (5-5) is a very poor approximation and gives a mean free path too large by more than an order of magnitude. The reason for this is that electrostatic forces decrease much more slowly with increasing distance than do the forces between neutral atoms. As a result, when two charged particles pass by at a distance large compared with p_0, the deflection χ is not negligible, and the number of such "distant encounters" is so great that their effect outweighs that of the close encounters.

More specifically, the equations above show that when p much exceeds p_0, ψ approaches $\frac{1}{2}\pi - p_0/p$; the deflection χ then varies as $2p_0/p$, and is thus a very slowly varying function of p. If all encounters produced deflections in the same direction, the distant collisions would have an enormous effect, since the number of collisions with an impact parameter between p and $p + dp$ varies as $2\pi p dp$. Actually, the deflections will have random directions and will tend to cancel out. To analyze distant collisions a statistical theory is required, which treats the effect produced by many small random changes in velocity. The fundamental concepts required by such a theory are presented in the next section.

5.2 Diffusion Coefficients

Let us follow a particular particle as it moves through an ionized gas. We shall call this particle a "test particle." The test particle has a mass m, a charge Ze/c, and a velocity $\mathbf{w}$. The particles with which the test particles collide will be called "field particles," in accordance with the terminology of Chandrasekhar (6). For simplicity we shall assume that all the field particles have the same mass, m_f, and charge, $Z_f e/c$. The field particles may have any distribution of velocities.

As the test particle moves about, it will experience many deflections, mostly small. Since the addition of successive angular deflections is mathematically complicated, we shall consider the value of $\Delta\mathbf{w}$, the change of velocity of the test particle. Let us choose the z axis parallel to $\mathbf{w}$, and consider firstly the component of $\mathbf{\Delta w}$ along the x axis. If $(\Delta w_x)_j$ represents the change of this component in the jth encounter, then in N encounters the total change of w_x is given by

$$\begin{aligned} \Delta w_x &= (\Delta w_x)_1 + (\Delta w_x)_2 + \cdots (\Delta w_x)_N \\ &= \textstyle\sum_j (\Delta w_x)_j \end{aligned} \tag{5-6}$$

We assume that successive encounters are at random. It is then impossible to predict what the precise value of Δw_x will be. However, if we consider many test particles, each moving in the same initial direction with the same initial velocity, and each experiencing N encounters, it is possible to average Δw_x over all these particles. We shall denote such an average by $\overline{\Delta w_x}$. If the distribution of velocities within the gas is isotropic, $\overline{\Delta w_x}$ must evidently vanish from symmetry. However, $\overline{\Delta w_z}$ is not necessarily zero in this case.

The mean square value of Δw_x will not vanish. This average will contain terms $\overline{(\Delta w_x)_j^2}$ and also such terms as $\overline{(\Delta w_x)_j(\Delta w_x)_k}$. If $(\Delta w_x)_j$ in each collision is small, the second collisions the particles experience will produce the same average change as the first collisions. The N terms $\overline{(\Delta w_x)_j^2}$ are all equal, therefore. On the other hand, all the cross-product terms will vanish, when averaged over all the particles under consideration, since successive collisions are uncorrelated. Hence we have

$$\overline{(\Delta w_x)^2} = N\overline{(\Delta w_x)_j^2} \tag{5-7}$$

If we plot the velocity of each test particle on a diagram, as in Figure 5.2, for the group of particles under consideration, all the points will initially coincide as in plot *a*. After N encounters, the points will spread out on the diagram, as shown in plots *b* and *c*. The dispersion of the points will increase as $N^{\frac{1}{2}}$, but may have different values in different directions. The center of gravity may move by an amount proportional to N. It is assumed, of course, that the test particles represented by the

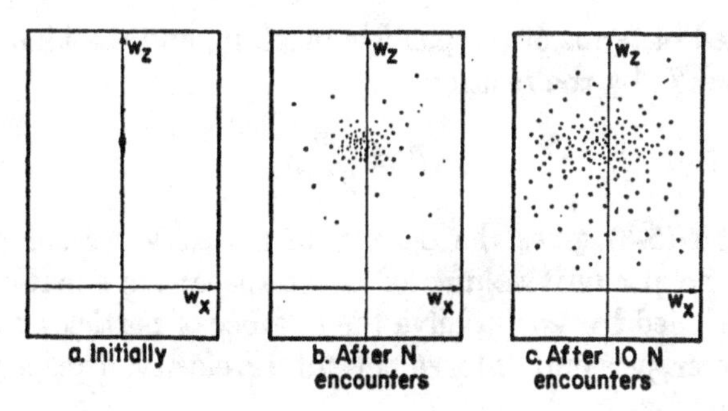

Figure 5.2. Diffusion of test particles in velocity space.

points in the diagram are but a small fraction of the total number of particles present.

We see that as a result of successive encounters the velocity distribution of a group of charged particles is progressively broadened. The effect is analogous to diffusion of particles in an ordinary gas, and we may regard the encounters as producing diffusion in velocity space.

To measure the rate of diffusion in the w_x direction, we set N in equation (5-7) equal to the average number of encounters in one second, including encounters of all types. The resultant value of $\overline{(\Delta w_x)^2}$, measuring the increase of velocity dispersion of a group of particles per second, may then be denoted by $\langle(\Delta w_x)^2\rangle$. This quantity is called a "diffusion coefficient." The corresponding quantities $\langle\Delta w_x\rangle$ and $\langle(\Delta w_x)(\Delta w_y)\rangle$ are also diffusion coefficients, which vanish from symmetry if the field particles have an isotropic velocity distribution. Diffusion coefficients may also be defined for other directions.

The diffusion coefficients may be evaluated, in principle, for any velocity distribution of the field particles. Of primary importance is the velocity distribution in kinetic equilibrium, the Maxwell-Boltzmann distribution, which may be written

$$f^{(0)}(w) = \frac{nl^3}{\pi^{3/2}} e^{-l^2w^2} \tag{5-8}$$

where n is the particle density of the particles in question; l

is defined in terms of the particle mass, m, and the kinetic temperature, T, by the relation

$$l^2 = \frac{m}{2kT} \tag{5-9}$$

Equation (5-8) gives the density of particles per cubic centimeter and per unit volume of phase space; the equation must be multiplied by $4\pi w^2$ to give the number of particles per cubic centimeter per unit interval of total velocity, irrespective of direction.

When the distribution function of the field particles is assumed to obey equation (5-8) only three independent diffusion coefficients need be considered, $\langle \Delta w_{\parallel} \rangle$, $\langle (\Delta w_{\parallel})^2 \rangle$, and $\langle (\Delta w_{\perp})^2 \rangle$. The first of these equals $\langle \Delta w_z \rangle$, in the coordinate system used in this section. This quantity, which is generally negative, represents the rate at which moving test particles are slowed down by interactions with the field particles. Chandrasekhar (7) has called this diffusion coefficient the "coefficient of dynamical friction." The quantity $\langle (\Delta w_{\perp})^2 \rangle$, or $\langle (\Delta w_z)^2 \rangle$ in the present notation, represents the rate of increase of $(\Delta w)^2$ in the direction parallel to the original motion of the test particles. The corresponding quantity $\langle (\Delta w_{\perp})^2 \rangle$ represents the rate of increase in the perpendicular direction, and equals twice $\langle (\Delta w_x)^2 \rangle$ or $\langle (\Delta w_y)^2 \rangle$.

To illustrate the principles involved we shall evaluate $\langle (\Delta w_{\perp})^2 \rangle$ in the simple case where m_f, the mass of the field particles, is much greater than m, the test-particle mass, and w_f may be set equal to zero. Usually in evaluating a diffusion coefficient we must integrate over the velocity distribution of the field particles, but in the present simple case this integration drops out.

When w_f vanishes we obtain, from equation (5-1),

$$(\Delta w_{\perp})^2 = w^2 \sin^2 \chi = 4w^2 \sin^2 \psi \cos^2 \psi \tag{5-10}$$

If we eliminate ψ by means of equations (5-2) and (5-4) we have

$$(\Delta w_{\perp})^2 = \frac{4w^2(p/p_0)^2}{\{1 + (p/p_0)^2\}^2} \tag{5-11}$$

If we average over all the test particles, the number of encounters with an impact parameter between p and $p + dp$, per second, will be $2\pi pwn_f dp$. Hence we have

$$\langle(\Delta w_\perp)^2\rangle = 8\pi n_f w^3 p_0^2 \int_0^{p_m/p_0} \frac{x^3 dx}{(1 + x^2)^2} \tag{5-12}$$

The integral diverges at infinity. If we arbitrarily cut off the integration at a value of p/p_0 equal to some constant p_m/p_0, assumed large compared to unity, we have, approximately

$$\langle(\Delta w_\perp)^2\rangle = 8\pi n_f w^3 p_0^2 \ln (p_m/p_0) \tag{5-13}$$

It remains to evaluate the cut-off distance p_m. One might expect that for p_m one should take the mean distance between the field particles, about equal to $n_f^{-1/3}$; at greater distances many particles will be passing by simultaneously, and equation (5-1) is no longer applicable. However, it has been shown by Cohen, Spitzer, and Routly (9) that in certain simplified cases the deflections produced by field particles passing by at distances greater than $n_f^{-1/3}$ are correctly given by the formulas derived from two-body encounters. The physical explanation is that statistical fluctuations of the charge density move by at the same mean speed as the passing field particles, and these fluctuations produce deflections that must be taken in account. As shown by Pines and Bohm (27), over distances greater than h the fluctuations of charged particle density are no longer random, but are reduced by macroscopic electrical forces. We shall, therefore, set p_m equal to h, as defined in equation (2-3). More detailed treatments by Rand (28) and others give essentially this same result. Since the logarithmic term changes slowly, we shall usually take the mean value of p_0 for all the test particles, replacing mw^2 by $3kT$. We then obtain

$$\Lambda \equiv \overline{h/p_0} = \frac{3}{2ZZ_f e^3}\left(\frac{k^3T^3}{\pi n_e}\right)^{1/2} \tag{5-14}$$

Shielding by positive ions is neglected in this equation.

When the electron temperature exceeds about 4×10^5 degrees K, Λ must be somewhat reduced below the values obtained from equation (5-14), because of quantum-mechanical effects. An electron wave passing through a circular aperture of radius p will be spread out by diffraction through an angle $\lambda/2\pi p$, where λ is the electron wavelength. If this deflection exceeds the classical deflection $2p_0/p$, then the previous equations must be modified; in accordance with the results of Marshak (24) the only change needed is to reduce Λ by the ratio

Table 5.1. Values of ln Λ

T, °K	Electron Density n_e, cm^{-3}								
	1	10^3	10^6	10^9	10^{12}	10^{15}	10^{18}	10^{21}	10^{24}
10^2	16.3	12.8	9.43	5.97					
10^3	19.7	16.3	12.8	9.43	5.97				
10^4	23.2	19.7	16.3	12.8	9.43	5.97			
10^5	26.7	23.2	19.7	16.3	12.8	9.43	5.97		
10^6	29.7	26.3	22.8	19.3	15.9	12.4	8.96	5.54	
10^7	32.0	28.5	25.1	21.6	18.1	14.7	11.2	7.85	4.39
10^8	34.3	30.9	27.4	24.0	20.5	17.0	13.6	10.1	6.69

$2\alpha c/w$, where α is the fine structure constant, equal to 1/137. Effectively the value of Λ given in equation (5-14) must be multiplied by $(4.2 \times 10^5/T)^{1/2}$, whenever T exceeds 4.2×10^5 degrees K. For collisions between positive ions the correction may be ignored. Values of ln Λ for an electron-proton gas are given in Table 5.1. For high densities and low temperatures the theory breaks down, and no values of ln Λ are given.

Detailed computations of the diffusion coefficients have been carried out by Chandrasekhar (6, 7), taking into account the motion of the center of mass of the colliding particles. The resultant formulas for the three diffusion coefficients are

$$\langle \Delta w_{\parallel} \rangle = -A_D l_f^2 \left(1 + \frac{m}{m_f}\right) G(l_f w) \qquad (5\text{-}15)$$

$$\langle(\Delta w_{\parallel})^2\rangle = \frac{A_D}{w} G(l_f w) \tag{5-16}$$

$$\langle(\Delta w_{\perp})^2\rangle = \frac{A_D}{w} \{\Phi(l_f w) - G(l_f w)\} \tag{5-17}$$

where the "diffusion constant," A_D, is defined by

$$A_D = \frac{8\pi e^4 n_f Z^2 Z_f^2 \ln \Lambda}{m^2} \tag{5-18}$$

$\Phi(x)$ is the usual error function,

$$\Phi(x) = \frac{2}{\pi^{1/2}} \int_0^x e^{-y^2} dy \tag{5-19}$$

and the function $G(x)$ is defined in terms of $\Phi(x)$ by the relationship

$$G(x) = \frac{\Phi(x) - x\Phi'(x)}{2x^2} \tag{5-20}$$

Values of G and $\Phi - G$ are given in Table 5.2. It is evident that as m_f approaches infinity, l_f also becomes infinite, according to equation (5-9), $\Phi(l_f w) - G(l_f w)$ approaches unity, and equation (5-17) reduces to (5-13). From equation (5-9) we see that the quantity $l_f w$ occurring in these equations is simply the ratio of w to the root mean square two-dimensional velocity of the field particles.

In the derivation of equations (5-15) through (5-17), only those terms proportional to $\ln \Lambda$ have been retained. These have been called "dominant terms" by Chandrasekhar. It is evident from the values in Table 5.1 that $\ln \Lambda$, which is usually about equal to the ratio of dominant to non-dominant terms, is not very great, in general, amounting to one or two orders of magnitude. In certain special cases the non-dominant terms may actually exceed the dominant ones. In the exact formula for $\langle(\Delta w_{\parallel})^2\rangle$, for example, the terms neglected in equation (5-16) actually exceed those retained whenever $l_f^2 w^2$ is greater than about $\ln \Lambda$. Moreover, when $l_f w$ is large, the true value of the coefficient $\langle(\Delta w_{\parallel})^2\rangle$ no longer gives the increase of velocity dispersion parallel to the initial velocity. The rate of increase of dispersion, parallel to the particle motion, is given rigorously by $\overline{(\Delta w_{\parallel})_j^2} - \overline{(\Delta w_{\parallel})_j}^2$, summed over all encounters j per unit time. The sum of $\overline{(\Delta w_{\parallel})_j}^2$ is non-dominant and has been neglected, but when $l_f w$ is large, this approximation is no longer justified. Evidently, when the velocities of the

Table 5.2. Values of $G(x)$ and $\Phi(x) - G(x)$

x	0.0	0.1	0.2	0.3	0.4	0.5	0.6	0.7	0.8	0.9
$G(x)$	0	.037	.073	.107	.137	.162	.183	.198	.208	.213
$\Phi(x) - G(x)$	0	.075	.149	.221	.292	.358	.421	.480	.534	.584
x	1.0	1.1	1.2	1.3	1.4	1.5	1.6	1.7	1.8	1.9
$G(x)$	.214	.211	.205	.196	.186	.175	.163	.152	.140	.129
$\Phi(x) - G(x)$	.629	.669	.706	.738	.766	.791	.813	.832	.849	.863
x	2.0	2.5	3.0	3.5	4.0	5.0	6.0	7.0	8.0	10.0
$G(x)$	.119	.080	.056	.041	.031	.020	.014	.010	.008	.005
$\Phi(x) - G(x)$	.876	.920	.944	.959	.969	.980	.986	.990	.992	.995

test particles much exceed those of the field particles, equation (5-16) may be inaccurate. For the other two diffusion coefficients the non-dominant terms are generally less important and may usually be ignored.

A number of interesting physical conclusions may be drawn directly from equations (5-15) through (5-17). When w vanishes, $\langle \Delta w_{\parallel} \rangle$ must also vanish, and the distinction between displacements parallel and perpendicular to w must disappear. However, since there are two directions included in $\langle (\Delta w_{\perp})^2 \rangle$ and only one in $\langle (\Delta w_{\parallel})^2 \rangle$, the former quantity is twice the latter, for vanishing w. In the other limiting case, when w exceeds the random velocity of the field particles, $\langle (\Delta w_{\perp})^2 \rangle$ varies as A_D/w, while $\langle (\Delta w_{\parallel})^2 \rangle$ is less by a factor $1/(2l_f^2 w^2)$. When a group of test particles is moving more rapidly than the field particles, the diffusion of the corresponding points in velocity space is mostly sideways, i.e., perpendicular to the original velocity.

However, when $l_f^2 w^2$ increases above $\ln \Lambda$, the ratio of $\langle(\Delta w_{\parallel})^2\rangle$ to $\langle(\Delta w_{\perp})^2\rangle$ no longer decreases so rapidly, as equation (5-16) is no longer a valid approximation.

As m increases, with m_f, l_f and w all held constant, $\langle\Delta w_{\parallel}\rangle$ becomes progressively greater compared to the other two coefficients, although the decrease of A_D makes all the coefficients small. As we shall see in the next section, for test particles of large mass, moving at appreciable velocities through particles of smaller mass, the slowing down produced by dynamical friction is much more important than the increase in velocity dispersion produced by the other two coefficients.

5.3 Relaxation Times

The "time of relaxation" is a term frequently used to denote the time in which collisions produce a large alteration in some original velocity distribution. In view of the complexity of the various situations possible, this concept is not a very clearly defined one. It is possible to introduce certain times, which are defined in terms of the diffusion coefficients and which measure the rate at which velocity distributions will alter in certain ways, as a result of encounters between charged particles.

The time between collisions, or the reciprocal of the collision frequency, may be regarded as the time in which deflections gradually deflect the test particles by 90°. More precisely, we may define a "deflection time" t_D, by the equation

$$\langle(\Delta w_{\perp})^2\rangle t_D = w^2 \tag{5-21}$$

If the diffusion coefficient $\langle(\Delta w_{\perp})^2\rangle$ remained constant as the test particles gradually diffused over velocity space, then in a time t_D the cumulative root mean square value of $\sin \chi$, where χ is the deflection angle, would amount to unity, corresponding to a mean deflection of approximately 90°. Only in special cases—electrons moving through a field of nearly motionless heavy ions, for example—is this simple interpretation possible.

If we substitute from equation (5-17) for the diffusion coefficient, we obtain

$$t_D = \frac{w^3}{A_D\{\Phi(l_f w) - G(l_f w)\}} = \frac{1}{8\pi n_f w p_0^2(\Phi - G)\ln\Lambda} \tag{5-22}$$

where p_0 is defined in equation (5-4). When m_f is large and $l_f w$ is, consequently, large, Φ equals unity and G vanishes. The value of t_D given by equation (5-22) is less by a factor $1/(8 \ln \Lambda)$ than was found in equation (5-5) from a consideration of single encounters. A glance at Table 5.1 shows that this factor may amount to less than 0.01.

Table 5.3. Ratio of Relaxation Times

$l_f w$	0.0	0.2	0.4	0.6	0.8	1.0	1.2	1.4
t_D/t_E	2.00	1.99	1.88	1.74	1.56	1.36	1.16	0.97
$l_f w$	1.6	1.8	2.0	2.5	3.0	3.5	4.0	5.0
t_D/t_E	0.80	0.66	0.54	0.35	0.24	0.17	0.13	0.082

An energy exchange time t_E may also be defined by the relation

$$\langle(\Delta E)^2\rangle t_E = E^2 \tag{5-23}$$

The change of energy, ΔE, in a single encounter is given by

$$\Delta E = \frac{m}{2}\{2w\Delta w_{\parallel} + (\Delta w_{\parallel})^2 + (\Delta w_{\perp})^2\} \tag{5-24}$$

If only dominant terms are retained, $\langle(\Delta E)^2\rangle$ equals $m^2w^2\langle(\Delta w_{\parallel})^2\rangle$. Equation (5-23) then yields

$$t_E = \frac{w^3}{4A_D G(l_f w)} \tag{5-25}$$

Values of the ratio t_D/t_E, obtained on dividing equation (5-22) by equation (5-25), are given in Table 5.3. For the higher values of $l_f w$ these ratios are not too reliable, as the non-dominant terms begin to affect t_E. Chandrasekhar (5) has considered the contribution of non-dominant terms to $\langle(\Delta E)^2\rangle$, without, however, considering the additional term needed to give the true rate at which the dispersion of E increases. He finds that as $l_f w$ becomes very large, $G \ln \Lambda$ in equation (5-25) must be replaced by $0.5(1 + m_f/m)^{-2}$. It is physically clear that t_E should approach infinity as m_f/m approaches infinity, and the velocities of the field particles vanish. In this situation the test particles move in a fixed potential field, and their energy remains rigorously constant.

An important special case is provided by a group of particles interacting with themselves. If we consider particles whose velocity has the root mean square value for the group, then lw equals $(1.5)^{1/2}$, or 1.225, and t_D/t_E is 1.14. Thus t_D is about equal to t_E in this case and provides a measure both of the time required to reduce substantially any lack of isotropy in the velocity distribution and also of the time required for the distribution of kinetic energies to approach the Maxwellian distribution. We shall devote this particular value of t_D, with lw equal to $(1.5)^{1/2}$, by the symbol t_c, which we may call the "self-collision time" for a group of particles interacting with each other. From equation (5-22) we obtain, inserting numerical values,

$$t_c = \frac{m^{1/2}(3kT)^{3/2}}{8 \times 0.714\pi n e^4 Z^4 \ln \Lambda} = \frac{11.4 A^{1/2} T^{3/2}}{n Z^4 \ln \Lambda} \text{ sec} \tag{5-26}$$

where T is in degrees K. We have let m equal Am_0, where m_0 is the mass of unit atomic weight. For electrons A is 1/1836, and to obtain t_{ce}, the self-collision time for electrons, $11.4A^{1/2}/Z^4$ in equation (5-26) must be replaced by 0.266. Evidently t_{ce} is less than the self-collision time for protons by 1/43, the inverse ratio of the velocities. Thus the mean free path for electrons colliding with electrons is exactly equal to the mean free path for protons colliding with protons, provided that electrons and protons have the same kinetic temperature.

The value of t_c given in equation (5-26) determines the value of γ to be used for computing T in an adiabatic compression (see equation 1-33). If the compression is slow compared to t_c, γ equals 5/3. For compression more rapid than t_c, γ will equal 2 or 3, depending on whether the compression is perpendicular to the lines of magnetic force (two-dimensional compression) or parallel to **B** (one-dimensional compression).

It is of interest to note from equations (5-18), (5-22), and (5-25) that the times t_D and t_E do not involve m_f, except indirectly through l_f, and depend on m, the test-particle mass, directly as m^2. For the distant encounters which are considered here the particles move in nearly rectilinear paths, and the acceleration of the field particles during an encounter has a relatively small effect on the test particles.

The previous considerations relate to the increase of dispersion in the velocities and energies of the test particles. Under some conditions we are interested primarily in the rate at which the mean velocity of the test particles is decreased by encounters. For this purpose, we may introduce a "slowing-down" time t_s, defined by the relation

$$\langle \Delta w_{\parallel} \rangle t_s = -w \tag{5-27}$$

Evidently the mean velocity of the test particles decreases at the rate w/t_s. If m, the test-particle mass, much exceeds m_f, then the mean kinetic energy, W, of the test particles will decrease at the rate $-2W/t_s$. For m much less than m_f, however, t_s represents the effect of deflections rather than of energy losses.

From equation (5-15), we find

$$t_s = \frac{w}{(1 + m/m_f)A_D l_f^2 G(l_f w)} \tag{5-28}$$

Equation (5-28) has two important limiting cases. If w much exceeds the root mean square velocity of the field particles, then $l_f w$ is large, $l_f^2 G(l_f w)$ equals $1/2w^2$ and t_s varies as w^3. On the other hand, if w is much less than the random velocity of the field particles, $l_f w$ is small, and t_s approaches a constant value, given by

$$
\begin{aligned}
t_s &= \frac{3\pi^{1/2}}{2(1 + m/m_f)A_D l_f^3} \\
&= 11.7 \frac{A^2 T_f^{3/2}}{(A + A_f)A_f^{1/2} n_f Z^2 Z_f^2 \ln \Lambda} \text{ sec}
\end{aligned}
\tag{5-29}
$$

When equation (5-29) is applicable, the mean velocity of the test particles will approach zero exponentially, with a time constant t_s. The decay rate of the kinetic energy, W, may be completed by taking mean values in equation (5-24) and using equations (5-15) through (5-17). If m/m_f is large, the decay time for W equals $t_s/2$.

Detailed numerical calculations on the energy distribution of deuterons injected into a deuterium plasma have been carried out by Kranzer (19). His results show that, in accordance with equation (5-28), the thermalization time exceeds t_c for the plasma deuterons by about a factor $(A_i/A_e)^{1/2}$ if the speed, w, of the injected ions exceeds the random velocity of the plasma ions by about $(A_i/A_e)^{1/6}$, and electrons moving faster than the injected deuterons provide the retardation. If w exceeds the electron thermal velocity, the thermalization time is even greater, and varies as w^3.

Finally, we consider the rate at which equipartition of energy is established between two groups of particles. Let us suppose that the test particles and the field particles both have Maxwellian velocity distributions, but with different kinetic temperatures T and T_f. If we use equation (5-24) to find $\langle \Delta E \rangle$, and average this over a Maxwellian velocity distribution for the test particles we find the result obtained by Spitzer (34),

$$
\frac{dT}{dt} = \frac{T_f - T}{t_{eq}} \tag{5-30}
$$

where t_{eq}, the time of equipartition, is given by

$$
\begin{aligned}
t_{eq} &= \frac{3 m m_f k^{3/2}}{8(2\pi)^{1/2} n_f Z^2 Z_f^2 e^4 \ln \Lambda} \left(\frac{T}{m} + \frac{T_f}{m_f} \right)^{3/2} \\
&= 5.87 \frac{A A_f}{n_f Z^2 Z_f^2 \ln \Lambda} \left(\frac{T}{A} + \frac{T_f}{A_f} \right)^{3/2} \text{ sec}
\end{aligned}
\tag{5-31}
$$

Equation (5-31) indicates that if the mean square relative velocity of the particles, which is proportional to $(T/m) + (T_f/m_f)$, does not change appreciably, then t_{eq} is constant. If also T_f is constant, departures from equipartition will decrease exponentially with the time constant t_{eq}. If $nT + n_fT_f$ is constant, the time constant for the approach to equilibrium will be $t_{eq} \times (1 + n/n_f)^{-1}$.

We are now in a position to discuss what happens in a proton-electron gas, for example, when the velocity distribution is originally arbitrary. We assume that the mean kinetic energies of electrons and protons are of the same order of magnitude. Collisions of electrons with protons will deflect the electrons and lead to an isotropic velocity distribution, but will not change appreciably the distribution of electron kinetic energies; t_D/t_E is small for l_f/w large. Electron-electron collisions will gradually establish a Maxwellian velocity distribution for the electrons, while the proton-proton collisions will yield a corresponding velocity distribution for the protons, but at a kinetic temperature that may differ from the electron temperature. The electrons, because of their greater velocity, will come to a Maxwellian equilibrium more rapidly than will the protons. We have already seen that t_c for electrons is less than for protons by the square root of A_e/A_p, the mass ratio of electron to protons, or by a factor of 1/43. Finally equipartition between electrons and protons is established by electron-proton collisions. These are relatively ineffective in exchanging energy; hence t_{eq} contains a factor $1/A_e^{1/2}$. If the small difference in numerical factors between equations (5-26) and (5-31), is neglected, t_{eq} is 43 times the collision time, t_c, for protons, and 1836 times the corresponding collision time for electrons.

5.4 Electrical Resistivity

The electrical resistivity, η, has been defined in equation (2-13) by the ratio of $\mathbf{P}_{ei}$, the rate at which electrons in a unit volume gain momentum by impact with positive ions, to $\mathbf{j}$, the current density. With this definition, η is directly related to the heating effect produced by the flow of current through a resistive medium. The power dissipated into heat per unit volume equals the force on the electrons, resulting from ion impact, times the mean drift velocity of electrons relative to the ions. The first of these two quantities is simply $\mathbf{P}_{ei}$, while the latter is $c\mathbf{j}/ne$ (see equation 2-9). Use of the definition of η given in equation (2-13) gives directly that the rate of ohmic, or Joule, heating is ηj^2. Thus the coefficient η used in the preceding chapters has always a simple physical meaning in terms of the dissipation of energy by an electric current.

The more conventional procedure is to define η as the ratio of $\mathbf{E}$ to $\mathbf{j}$. In view of the complicated interrelationship between $\mathbf{E}$ and $\mathbf{j}$ when a magnetic field is present, the usual definition seems unsuitable. Nevertheless, we have seen that in many situations, Ohm's law is obeyed, with η defined in terms of $\mathbf{P}_{ei}$. In such situations, it is immaterial whether η is computed from P_{ei}/j or from E/j, since the results will be the same. In particular, when no magnetic field is present the two definitions are precisely identical. As we shall see below, however, the exact computation of η transverse to a magnetic field is carried through much more directly from equation (2-13) than from Ohm's law.

To compute η in order of magnitude we may use the most elementary form of the kinetic theory in which each collision is assumed to produce a large deflection. The momentum gained by an electron in each collision is then $m_e(\mathbf{v}_i - \mathbf{v}_e)$, on the average, where $\mathbf{v}_e$ and $\mathbf{v}_i$ are the macroscopic velocities of electrons and positive ions. The number of such collisions per cubic centimeter may be set equal to νn_e, where ν is the collision frequency for each electron. The current density is $n_e e(\mathbf{v}_i - \mathbf{v}_e)/c$. Combining these quantities in equation (2-13), we obtain

$$\eta = \frac{m_e c^2 \nu}{n_e e^2} \tag{5-32}$$

This familiar elementary equation for η is only approximate. In a fully ionized gas there is some uncertainty as to the appropriate value of ν to use in equation (5-32). The collision frequency should equal $1/t_D$; according to equation (5-22), t_D varies about as w^3, for $l_f w$ large, and thus varies enormously over the range of electron velocities present.

To compute the current with any precision in this situation, one must use the Boltzmann equation for f, the density of electrons in phase space. This equation is discussed briefly in the Appendix. When a steady current flows in a homogeneous medium, $\partial f/\partial t$ and $\partial f/\partial x_i$ vanish, and the basic equation (6-1) becomes

$$-\frac{\partial f(\mathbf{w})}{\partial w}\frac{eE\cos\theta}{m_e c} = \left(\frac{\partial f(\mathbf{w})}{\partial t}\right)_{\text{coll}} \tag{5-33}$$

where θ is the angle between $\mathbf{w}$ and the electric field $\mathbf{E}$. The quantity $(\partial f(\mathbf{w})/\partial t)_{\text{coll}}$ has been evaluated generally by Rosenbluth, MacDonald, and Judd (29). The solution of the resultant equation is a matter of some complexity.

A relatively simple situation is provided by the so-called Lorentz gas, a hypothetical fully ionized gas in which the electrons do not interact with each other, and all the positive ions are at rest. In such a gas the diffusion coefficients are very simple, and f may be obtained accurately if the electric field $\mathbf{E}$ is sufficiently small. The electrical resistivity of such an ideal Lorentz gas, which we denote by η_L, is

$$\eta_L = \frac{\pi^{3/2} m_e^{1/2} Z e^2 c^2 \ln \Lambda}{2(2kT)^{3/2}} \tag{5-34}$$

where Z is the ionic charge. Inserting numerical values, we find

$$\begin{aligned} \eta_L &= 3.80 \times 10^{12} \frac{Z \ln \Lambda}{T^{3/2}} \text{ e.m.u.} \\ &= 3.80 \times 10^{3} \frac{Z \ln \Lambda}{T^{3/2}} \text{ ohm-cm} \end{aligned} \tag{5-35}$$

To obtain an accurate expression for η in an ionized gas, electron-electron encounters must be taken into account. Several investigations along this line have been carried out. The general theory of Chapman and Cowling (8) has been applied to the problem by Cowling (10) and, in fuller detail, by Landshoff (20). The corresponding analysis in terms of diffusion coefficients has been worked out for this case by Cohen, Spitzer, and Routly (9), and final values for η obtained by Spitzer and Härm (36). These various analyses are in full agreement, and their results may be expressed in the form

$$\eta = \frac{\eta_L}{\gamma_E} \tag{5-36}$$

where values of γ_E, which depends on the ionic charge Z, are given in Table 5.4. Thus in the important case Z equal to 1,

Table 5.4. Ratio of Conductivity to that in a Lorentz Gas

Ionic Charge Z	1	2	4	16	∞
γ_E	0.582	0.683	0.785	0.923	1.000

we have

$$\begin{aligned}\eta &= 6.53 \times 10^{12} \frac{\ln \Lambda}{T^{3/2}} \text{ e.m.u.} \\ &= 6.53 \times 10^{8} \frac{\ln \Lambda}{T^{3/2}} \text{ ohm-cm}\end{aligned} \tag{5-37}$$

Observational confirmation of this result has been obtained by Lin, Resler, and Kantrowitz (21) and by Maecker, Peters, and Schenk (23). The agreement obtained with equation (5-37) must be regarded as somewhat fortuitous, however, in view of the small values of ln Λ, between 3 and 6, in these particular experiments.

The results obtained above are based on the assumption that E is sufficiently small so that the potential energy gained across one mean free path is negligible compared to kT. Since the mean free path increases about as w^4, it is clear that for sufficiently great velocities this basic assumption must break down; the energy gained in a mean free path will become greater and greater as w increases. When this computed energy gain per mean free path is comparable with the kinetic energy, there is an appreciable probability that the particle will accelerate indefinitely along the electric field (if we assume an infinite plasma and a constant E). The probability of a velocity change, as a result of encounters with other charged particles, decreases so rapidly with increasing E that once such a particle exceeds a certain critical velocity the influence of any further collisions is small. Such a continuously accelerating particle is called a "runaway."

The velocity above which electrons tend to run away can be computed from the condition that $-\langle \Delta w_{\parallel} \rangle$ be less than the acceleration eE/mc. We consider an electron whose velocity, $\mathbf{w}$, is parallel to $\mathbf{E}$; we assume that w is much larger than the root mean square particle velocities of ions or electrons, so that $l_f^2 G(l_f w)$ in equation (5-15) can be replaced by $1/(2w^2)$. Under these conditions we obtain the following condition for electron runaway

$$\frac{m_e w^2}{2kT} > \frac{1}{\Gamma}\left(1 + \frac{2}{Z}\right) \tag{5-38}$$

where we have taken into account dynamical friction both with other electrons and with ions of charge Ze/c; the dimensionless parameter Γ is given by

$$\Gamma = \frac{kTE}{2\pi c Z e^3 n_e \ln \Lambda} = 6.61 \times 10^8 \frac{TE \text{ (volts/cm)}}{Z n_e \ln \Lambda} \tag{5-39}$$

The quantity Γ equals approximately the ratio of the mean electron drift velocity to the random thermal velocity, $(3kT/m_e)^{1/2}$; for a homogeneous Lorentz gas equations (2-9), (2-24), and (5-34) may be combined to show that this ratio equals Γ times $8(2/3\pi)^{1/2}$, or 3.69Γ. Evidently Γ must be much less than unity if equation (5-37) for η is to be valid.

The rate at which electrons diffuse up to high velocities and run completely away has been computed approximately by Dreicer (12) for the case $Z = 1$. His numerical results, obtained for Γ in the range from 0.3 to 0.03, may be represented accurately by the equation

$$P_r = \frac{24eEe^{-(8/\Gamma)^{1/2}}}{\Gamma c(2mkT)^{1/2}} = \frac{30.9}{t_{ce}} \times e^{-(8/\Gamma)^{1/2}} \qquad (5\text{-}40)$$

where $P_r dt$ is the fraction of electrons that run away during a time interval dt, and t_{ce} is the self-collision time for electrons, given in equation (5-26).

When Γ is very small electron runaway is unimportant and the linearized theory leading to the numerical values in Table 5.4 is valid. The distribution of current over electrons of different velocity under these conditions is of interest. Figure 5.3 shows different curves for j_w; $j_w dw$ is the contribution to the electric current by electrons of total velocity between w and $w + dw$. Essentially j_w is proportional to w times the excess number of electrons going one way, at the velocity w, over the corresponding number going the other way. All the different curves are normalized to the same area. For comparison, the dashed line shows the number of electrons per unit velocity interval according to the Maxwell-Boltzmann formula $w^2 f^{(0)}(w)$, where $f^{(0)}(w)$ is given in equation (5-8). Curve (c) is drawn for a Lorentz gas, in which j_w is proportional to $w^7 f^{(0)}(w)$. This curve is valid for a gas in which the positive ions have a very high nuclear charge, since in such a gas the electron-electron collisions are negligible. Curve (b) gives j_w for an electron-proton gas, based on results by Spitzer and Härm (36).

In the presence of a magnetic field transverse to an electric field these results are no longer valid, since the distribution of current over electrons of different velocities is altered. Let us consider a strong magnetic field, such that the radius of gyration a is much less than the mean free path between collisions; i.e., ω_{ce} much exceeds the collision frequency ν. Then the single-particle analysis in Chapter 1 is valid, and if the temperature is constant the current at a point P arises because there are more guiding centers on one side of P than on the other. Hence the contribution to j by electrons of velocity w will depend on the difference between the density of guiding centers on the two sides of P. The distance of these guiding centers from P is a, the radius of gyration, and since a varies linearly with w and the density varies linearly with distance, to a first approximation, the surplus of electrons going in one direction over electrons going the other way will be proportional to w. Since the electric current is itself proportional to the excess number of electrons, multiplied by their velocity, it follows that j_w will vary as w^2 times the number of particles per unit interval of total velocity, or as $w^4 f^{(0)}(w)$. This variation of j_w with w is shown by curve (a) in Figure 5.3.

For a given current the transfer of momentum from electrons to positive ions, which is the basis for the definition of η

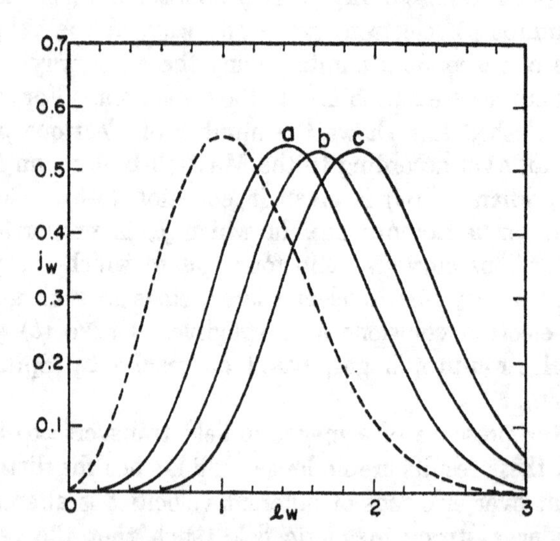

Figure 5.3. Contribution to current by electrons of different velocities. The dashed line represents the Maxwell-Boltzmann distribution; the other curves depict j_w, normalized to unity, under the following conditions: (a) transverse to a strong magnetic field, **B**; (b) electron-proton gas with **B** = 0; (c) an electron gas with ions of very high Z, **B** = 0.

in Section 2.2, will clearly depend on how the current is distributed over electrons of different velocity. If, for example, the current were entirely concentrated in electrons of very high velocity, the transfer of momentum would be very small, as the cross section for interaction decreases rapidly with increasing velocity. For a given variation of j_w with w, η may be computed directly without reference to E. The computations, given by Spitzer (35), lead to the result that for a current transverse to a strong magnetic field the resistivity found for a Lorentz gas must be divided by a factor γ_{EB} where

$$\gamma_{EB} = \frac{3\pi}{32} = 0.295 \tag{5-41}$$

If we combine equations (5-35) and (5-41), we find for $\eta_\perp$, the resistivity transverse to a strong magnetic field,

$$\begin{aligned} \eta_\perp &= 1.29 \times 10^{13} \frac{Z \ln \Lambda}{T^{3/2}} \text{ e.m.u.} \\ &= 1.29 \times 10^{4} \frac{Z \ln \Lambda}{T^{3/2}} \text{ ohm-cm} \end{aligned} \tag{5-42}$$

In the presence of transverse thermal gradients the effective resistivity is altered by the thermoelectric terms discussed below —see equation (5-49).

5.5 Thermal Conductivity and Viscosity

The same methods used for the evaluation of the electrical conductivity may also be applied to other transport coefficients. We consider here the thermal conductivity and also the viscosity.

In the presence of a temperature gradient, ∇T, not only will a flow of heat, $\mathbf{Q}$, appear, but an electric current $\mathbf{j}$ will also flow. The temperature gradient warps the velocity distribution and a net flow of electrons appears. Similarly, an electric field produces a flow of heat. In the absence of a magnetic field we may write, for a steady state

$$\mathbf{j} = \frac{1}{\eta}\mathbf{E} + \alpha\nabla T \tag{5-43}$$

$$\mathbf{Q} = -\beta\mathbf{E} - \mathcal{K}\nabla T \tag{5-44}$$

According to the thermodynamics of irreversible processes (14), these four coefficients are not independent, but are related by the expression

$$\beta = \alpha T + \frac{5kT}{2e\eta} \tag{5-45}$$

The thermoelectric effects represented in equations (5-43) and (5-44) act to reduce the effective coefficient of thermal conductivity. In a steady state no current can flow in the direction of the temperature gradient, as a current divergence would result, and electric fields would rise rapidly without limit. What happens is that a secondary electric field is produced such that the current produced by the temperature gradient is cancelled. This secondary electric field, in turn, acts to reduce the flow of heat. The effective coefficient of conductivity is reduced to $\epsilon\mathcal{K}$, where

$$\epsilon = 1 - \frac{\beta\alpha\eta}{\mathcal{K}} \tag{5-46}$$

For a Lorentz gas the value of $\mathcal{K}$ is

$$\begin{aligned}\mathcal{K}_L &= 20\left(\frac{2}{\pi}\right)^{3/2}\frac{(kT)^{5/2}k}{m_e^{1/2}e^4Z\ln\Lambda} \\ &= 4.67\times10^{-12}\frac{T^{5/2}}{Z\ln\Lambda}\frac{\text{cal}}{\text{sec deg cm}}\end{aligned} \tag{5-47}$$

while ϵ is 0.40. For an actual gas $\mathcal{K}$ becomes

$$\mathcal{K} = \delta_T\mathcal{K}_L \tag{5-48}$$

Values of δ_T and ϵ are given in Table 5.5 for different values of Z, taken again from Spitzer and Härm (36), who give detailed results for α and β also. With these results it is readily

Table 5.5. Values of δ_T and ϵ

Z	1	2	4	16	∞
δ_T	0.225	0.356	0.513	0.791	1.000
ϵ	0.419	0.410	0.401	0.396	0.400

verified that equation (5-45) is satisfied to about one part in a thousand, a useful check on the numerical accuracy of the work.

A strong magnetic field will reduce the transverse heat flow, $\mathbf{Q}_\perp$. If we assume a quasi-steady state, we obtain in the limit of weak collisions (collision frequency ν much less than ω_{ci}) the following coupled equations:

$$\mathbf{j}_\perp = \frac{1}{\eta_\perp}\mathbf{E}' + \lambda\mathbf{B}\times\nabla T \tag{5-49}$$

$$\mathbf{Q}_\perp = -\lambda T\mathbf{B}\times\mathbf{E}' - \mathcal{K}\nabla T \tag{5-50}$$

$$\mathbf{E}' = \mathbf{E} + \mathbf{v}\times\mathbf{B} - \frac{c}{en_e}\nabla p_i \tag{5-51}$$

and it is assumed that both ∇T and $\mathbf{E}'$ have no component parallel to $\mathbf{B}$. Electrons and ions are assumed to have local Maxwellian velocity distributions. The coefficient $\eta_\perp$ is given in equation (5-42), while the other two coefficients are given by

$$\lambda = \frac{3kn_e}{2B^2} \tag{5-52}$$

$$\mathcal{K}_\perp = \frac{8(\pi m_i k)^{1/2} n_i^2 Z^2 e^2 c^2 \ln\Lambda}{3B^2 T^{1/2}} = 3.54\times10^{-25}\,\frac{A_i^{1/2}Z^2 n_i^2 \ln\Lambda}{T^{1/2}B^2}\;\frac{\text{cal}}{\text{sec deg cm}} \tag{5-53}$$

These equations, with Z equal to unity, have been derived by Rosenbluth and Kaufman (30). The transverse thermal conductivity is entirely due to positive ions, in this approximation, because of their large mass and large radius of gyration. It is readily seen that the ratio of $\mathcal{K}_\perp$ to $\mathcal{K}_L$ varies as $1/(\omega_{ci}t_{ci})^2$, where

t_{ci} is the self-collision time for positive ions. Equation (5-53) is valid only if $\omega_{ci}t_{ci}$ is large; when $\omega_{ci}t_{ci}$ is comparable with unity this equation is not applicable, and $\mathcal{K}$ may be obtained from the general discussion by Braginskii (4).

The viscosity of a fully ionized gas has been analyzed by Braginskii (4). In the absence of a magnetic field the coefficient of viscosity is given by

$$\mu = \frac{0.406\, m_i^{1/2}(kT)^{5/2}}{Z^4e^4 \ln \Lambda} = 2.21 \times 10^{-15} \frac{T^{5/2}A_i^{1/2}}{Z^4 \ln \Lambda} \frac{\text{gm}}{\text{cm sec}} \tag{5-54}$$

where A_i is again the atomic weight of the positive ions. The viscosity is due primarily to positive ions; viscous stresses due to electrons are generally negligible.

When a magnetic field is present, the value of μ will depend both on the direction of the velocity and the direction along which the gradient of velocity is considered. To compute the stress parallel to **B** resulting from gradients of $v_{\parallel}$ in a direction along **B**, equation (5-54) may be used directly. When the velocity **v** is perpendicular to the magnetic field, the stress resulting from a gradient in a direction perpendicular both to **B** and **v** may be expressed in terms of the coefficient $\mu_{\perp}$. As noted in Chapter 2, the viscosity transverse to a strong magnetic field produces a drift of ions given by equation (2-44). For a gas in which the magnetic field is very strong ($\omega_{ci}t_{ci}$ much greater than unity), $\mu_{\perp}$ is given by

$$\begin{aligned}\mu_{\perp} &= \frac{2}{5}\left(\frac{\pi}{m_ikT_i}\right)^{1/2} \frac{Z^4e^4n_i^2 \ln \Lambda}{\omega_{ci}^2} \\ &= 2.68 \times 10^{-26} \frac{A_i^{3/2}Z^2n_i^2 \ln \Lambda}{T_i^{1/2}B^2} \frac{\text{gm}}{\text{cm sec}}\end{aligned} \tag{5-55}$$

a result first obtained by Simon (33). Evidently the ratio of $\mu_{\perp}$ to μ obtained from equations (5-54) and (5-55) is about $1/\omega_{ci}^2t_{ci}^2$. A somewhat different analysis by Longmire and Rosenbluth (22) obtains a result similar to equation (5-55), but with a numerical coefficient greater by 4/3; the origin of this discrepancy is not clear. As shown by Kaufman (18) the effective coefficient of viscosity for stresses due to gradients of $v_{\parallel}$ in a direction transverse to **B** is $4\mu_{\perp}$, when the magnetic field is very strong. Results for the various stresses in the general case, including values of $\omega_{ci}t_{ci}$ comparable with unity, have been given by Braginskii (4).

5.6 Radiation

While the discussion in the preceding sections is essentially classical, the microscopic interaction of photons with the individual charged particles of an ionized gas must in many situations be treated by quantum mechanics. The theory of radiation is beyond the scope of this tract and is adequately covered by Heitler (16), Bethe and Salpeter (3), and others. Because of the importance of these radiative processes we give here a brief summary of the results obtained for the restricted case of a fully ionized gas.

We treat three processes by which a photon can interact with a free electron. First, a photon can be scattered by the electron. Second, a photon can be emitted or absorbed by an electron in the presence of a heavy positive ion. Third, an electron in a magnetic field may absorb or emit a photon whose frequency is nearly equal to ω_{ce} or to a multiple of ω_{ce}.

a. Photon Scattering by Free Electrons. The total cross section of an electron for scattering a photon is given by

$$\sigma_{e} = \frac{8\pi}{3}\left(\frac{e^2}{m_e c^2}\right)^2 = 6.65 \times 10^{-25}\ \text{cm}^2 \qquad (5\text{-}56)$$

Equation (5-56) neglects relativistic effects. When the photon energy becomes comparable with the electron rest-mass, the Klein-Nishina formula (16) must be used.

The scattered radiation tends to be polarized. In addition, the photon frequency may be altered in the process. If the electron is at rest initially, then the decrease of wavelength $\Delta\lambda$ is given by $\lambda_0(1 - \cos\chi)$, where χ is the angular deflection of the photon and λ_0 is the familiar Compton wavelength

$$\lambda_0 = \frac{h}{m_e c} = 2.43 \times 10^{-2} \text{ angstrom} \tag{5-57}$$

If the electron is moving rapidly, $\Delta\lambda$ may have a large positive or negative value as a result of the Doppler effect.

b. Photon Emission in Electron-Ion Collisions. When an electron emits a light quantum, with the heavy ion absorbing the momentum, the electron must decrease its energy. If the electron remains free, such transitions are called free-free transitions, and the radiation emitted is called bremsstrahlung. For electrons all of the same kinetic energy, less than the ionization energy for the ion in question, the spectrum of emitted energy per unit frequency is reasonably flat (3) from the maximum frequency $m_e w^2/2h$ down to about a tenth of this frequency. If the electrons have a Maxwellian distribution of velocities, with a kinetic temperature T, the emitted intensity per unit frequency varies about as exp $(-h\nu/kT)$. Corresponding to the emission of radiation, absorption of a photon in a free-free transition is also possible. The absorption coefficient κ_ν, in cm^2/cm^3, is approximately

$$\begin{aligned} \kappa_\nu &= \frac{4}{3}\left(\frac{2\pi}{3kT}\right)^{1/2} \frac{n_e n_i Z^2 e^6}{hcm_e^{3/2}\nu^3} g_{ff} \\ &= 3.69 \times 10^8 \frac{Z^3 n_i^2}{T^{1/2}\nu^3} g_{ff} \text{ cm}^{-1} \end{aligned} \tag{5-58}$$

where we have again replaced n_e by Zn_i. Induced emissions reduce the effective absorption coefficient; to take these into account, the value found from equation (5-58) must be multiplied by $1 - \exp(-h\nu/kT)$.

The quantity g_{ff} appearing in equation (5-58) is a correction factor required for precise results. Its value is generally about one. For radio waves, however, with ν much less than kT/h but much greater than the plasma frequency, $\omega_p/2\pi$, we have the result by Elwert (13) and Scheuer (31)

$$g_{ff} = \frac{3^{1/2}V}{\pi c}\left\{\ln \frac{(2kT)^{3/2}}{\pi Z e^2 \nu m_e^{1/2}} - \frac{5\gamma}{2}\right\} \tag{5-59}$$

where V is the wave velocity and γ is Euler's constant, equal to 0.5772. For frequencies comparable with $\omega_p/2\pi$ approximate formulae for g_{ff} have been given by Oster (26).

The total amount of energy radiated in free-free transitions, per cm^3 per sec, will be denoted by ϵ_{ff}. For a Maxwellian distribution of velocities ϵ_{ff} is given by

$$\begin{aligned} \epsilon_{ff} &= \left(\frac{2\pi kT}{3m_e}\right)^{1/2} \frac{2^5\pi e^6}{3hm_e c^3} Z^2 n_e n_i \bar{g}_{ff} \\ &= 1.42 \times 10^{-27} Z^3 n_i^2 T^{1/2} \bar{g}_{ff} \frac{\text{ergs}}{\text{cm}^3\ \text{sec}} \end{aligned} \tag{5-60}$$

If g_{ff} and $\bar{g}_{ff}$ are set equal to unity, equations (5-58) and (5-60) give Kramer's semiclassical result. The Born approximation for ϵ_{ff} is obtained if we set

$$\bar{g}_{ff} = \frac{2(3)^{1/2}}{\pi} = 1.103 \tag{5-61}$$

Exact values of $\bar{g}_{ff}$ over a wide range in n_i and T are given by Greene (15).

It is also possible for the electrons to become captured, with the emission of radiation. Especially at the higher temperatures this recombination radiation is negligible compared to the bremsstrahlung, but it is important in determining the rate at which positive ions recapture electrons. An electron may be captured into any level of total quantum number n. For a hydrogenic nucleus, the energy $h\nu$ of the photon emitted will be $\frac{1}{2}m_e w^2 + h\nu_0/n^2$, where $-h\nu_0$ is the energy of the ground state. The cross section for capture in level n, which we denote by σ_{cn}, equals

$$\sigma_{cn} = A_r \frac{\nu_0}{\nu} \frac{h\nu_0}{\frac{1}{2}m_e w^2} \frac{g_{fn}}{n^3} \tag{5-62}$$

where the "recapture constant," A_r, is given by

$$A_r = \frac{2^4}{3^{3/2}} \frac{he^2}{m_e^2 c^3} = 2.11 \times 10^{-22}\ \text{cm}^2 \tag{5-63}$$

For captures in the ground state (n equal to one) the correction factor g_{fn} becomes

$$g_{f1} = 8\pi 3^{1/2}\frac{\nu_0}{\nu}\frac{\exp\left\{-4\left(\frac{\nu_0}{\nu-\nu_0}\right)^{1/2}\tan^{-1}\left(\frac{\nu-\nu_0}{\nu_0}\right)^{1/2}\right\}}{1-\exp\left\{-2\pi\left(\frac{\nu_0}{\nu-\nu_0}\right)^{1/2}\right\}} \tag{5-64}$$

From equation (5-64) it may be seen that g_{f1} equals 0.797 for an electron of small velocity (ν about equal to ν_0), and decreases as $(\nu_0/\nu)^{1/2}$ for ν much greater than ν_0.

The rate of electron disappearance by radiative recombination is given by the equation

$$\frac{dn_e}{dt} = -\alpha n_e n_i \tag{5-65}$$

where

$$\alpha = \sum_n \overline{\sigma_{cn} w} \tag{5-66}$$

the average being taken over all electron velocities, with the sum extending over all the bound states into which an electron can be captured. If we assume a Maxwellian distribution of velocities and set g_{fn} equal to one in equation (5-62), equation (5-66) gives

$$\alpha = 2A_r\left(\frac{2kT}{\pi m_e}\right)^{1/2}\beta\phi(\beta) = 2.07\times 10^{-11}Z^2T^{-1/2}\phi(\beta) \tag{5-67}$$

where

$$\beta = \frac{h\nu_0}{kT} = \frac{157{,}000^\circ Z^2}{T} \tag{5-68}$$

and

$$\phi(\beta) = \sum_{n=1}^{\infty}\frac{\beta}{n^3}e^{\beta/n^2}\left\{-Ei\left(-\frac{\beta}{n^2}\right)\right\} \tag{5-69}$$

The quantity $-Ei(-x)$ is the familiar exponential integral. Values of $\phi(\beta)$ are given in Table 5.6.

The measured values of α in laboratory plasmas are usually many orders of magnitude greater than the values given above. As shown theoretically by Bates (1), this apparent discrepancy may be attributed to a different type of recombination process,

Table 5.6. Values of $\phi(\beta)$

β	0.01	0.02	0.05	0.1	0.2	0.5
T/Z^2	1.6×10^7	7.9×10^6	3.2×10^6	1.6×10^6	7.9×10^5	3.2×10^5
$\phi(\beta)$	0.053	0.090	0.18	0.28	0.43	0.70
β	1.0	2.0	5	10	100	1000
T/Z^2	1.6×10^5	7.9×10^4	3.2×10^4	1.6×10^4	1.6×10^3	1.6×10^2
$\phi(\beta)$	0.96	1.26	1.69	2.02	3.2	4.3

in which a molecular ion captures an electron and dissociates into two neutral atoms. In addition, at relatively high electron densities and low temperatures three-body collisions between an ion and two electrons are responsible for a large increase in the recombination rate, according to D'Angelo (11) and Bates and Kingston (2). Hinnov and Hirschberg (17) have shown that the recombination observed in dense hydrogen and helium plasmas agrees with the rate computed for this process.

According to equation (5-63) the cross section for radiative capture of an electron by a proton is very small relative to the geometrical cross section of the H atom, amounting to only 2.1×10^{-21} cm^2 for capture of a 1-volt electron in the ground level. In marked contrast, the cross section for ionization of a neutral atom by an energetic electron is much greater, about 10^{-16} cm^2 for an electron of 100 volts energy impinging on an H atom. Detailed theoretical values for collisional cross sections are given in the thorough survey by Massey and Burhop (25), who also give detailed data on the multifarious processes which involve excitation, ionization, neutralization, combination, and dissociation of atoms and molecules, and which are evidently outside the scope of the present work.

c. Synchrotron Radiation. When a single charged particle gyrates in a magnetic field, radiation is emitted in all harmonics of the gyration frequency. Since this radiation has been observed from synchrotrons, it is commonly referred to as synchrotron radiation. The early analysis by Schwinger (32) has been extended by Trubnikov (37) to include motions parallel to **B**. If $\beta_\perp$ and $\beta_\parallel$ denote $w_\perp/c$ and $w_\parallel/c$, respectively, then $\epsilon_{sn}(\theta)$, the emissivity per unit solid angle at an angle θ to **B**, in the harmonic of order n, is given by

$$\epsilon_{sn}(\theta) = \frac{e^2\omega_c^2(1 - \beta^2)y^2}{2\pi c(1 - \beta_\parallel \cos\theta)} \times \left\{\left(\frac{\cos\theta - \beta_\parallel}{\beta_\perp \sin\theta}\right)^2 J_n^2\,(y\sin\theta) + J_n'^2\,(y\sin\theta)\right\} \tag{5-70}$$

The quantity ω_c is the angular cyclotron frequency at the rest mass, J_n is the usual Bessel function of order n, β^2 equals the sum of $\beta_\perp^2$ and $\beta_\parallel^2$, and

$$y = \frac{n\beta_\perp}{1 - \beta_\parallel \cos\theta} \tag{5-71}$$

The angular frequency, ω, of the radiation emitted in this direction is given by

$$\omega = \omega_c y(1 - \beta^2)^{1/2}/\beta_\perp \tag{5-72}$$

As shown by Schwinger, the total energy radiated in all directions, summed over all n, is given by

$$\epsilon_s = \frac{2e^2}{3c^3}\frac{1}{(1 - \beta^2)^2}\left(\frac{d\mathrm{v}}{dt}\right)^2 = 1.59 \times 10^{-15}\,\frac{B^2\beta_\perp^2}{1 - \beta^2}\,\frac{\text{ergs}}{\text{sec}} \tag{5-73}$$

For highly relativisitic velocities the radiation is most intense in the harmonic with n about equal to $\beta_\perp(1 - \beta^2)^{-3/2}$, falling off sharply at greater n. Intensities of the emitted radiation in this case, for the two degrees of polarization, have been given by Westfold (38). In a plasma at moderate temperatures reabsorption must be taken into account; if the velocity distribution is Maxwellian, the emitted intensity cannot exceed that from a black body at the kinetic temperature.

References

1. Bates, D. R., *Phys. Rev.*, **77**, 718 (1950); *Phys. Rev.*, **78**, 492 (1950).
2. Bates, D. R. and A. E. Kingston, *Nature*, **189**, 652 (1961).
3. Bethe, H. A. and E. E. Salpeter, *Encyclopedia of Physics (Handbuch der Physik)* XXXV, Section 4, 1957.
4. Braginskii, S. I., *J. Exptl. Theoret. Phys. (U.S.S.R.)*, **33**, 459 (1957); *Soviet Phys. JETP*, **6**, 358 (1958).
5. Chandrasekhar, S., *Astrophys. J.*, **93**, 285 (1941).
6. Chandrasekhar, S., *Principles of Stellar Dynamics*, University of Chicago Press, Chicago, 1942, Chapter 2 and Section 5.6.
7. Chandrasekhar, S., *Astrophys. J.*, **97**, 255 (1943).
8. Chapman, S. and T. G. Cowling, *The Mathematical Theory of Non-Uniform Gases*, Cambridge University Press, Cambridge, 1953.
9. Cohen, R. S., L. Spitzer, and P. McR. Routly, *Phys. Rev.*, **80**, 230 (1950).
10. Cowling, T. G., *Proc. Roy. Soc. (London)*, **A183**, 453 (1945).
11. D'Angelo, N., *Phys. Rev.*, **121**, 505–507 (1961).
12. Dreicer, H., *Phys. Rev.*, **117**, 329 (1960).
13. Elwert, G., *Z. Naturforsch.*, **3A**, 477 (1948).
14. deGroot, S. R., *Thermodynamics of Irreversible Processes*, Interscience, New York-London, 1961, Chapter VIII.
15. Greene, J., *Astrophys. J.*, **130**, 693 (1959).
16. Heitler, W., *Quantum Theory of Radiation*, Clarendon Press, Oxford, 1953.
17. Hinnov, E. and J. G. Hirschberg, *Phys. Rev.*, **125**, 795 (1962).
18. Kaufman, A., *La Théorie dez Gas Neutres et Ionizés*, Hermann et Cie, Paris, 1960, p. 319.
19. Kranzer, H. C., *Phys. Fluids*, **4**, 214 (1960).
20. Landshoff, R., *Phys. Rev.*, **76**, 904 (1949); *Phys. Rev.*, **82**, 442 (1951).
21. Lin, S., E. L. Resler, and A. Kantrowitz, *J. Appl. Phys.*, **26**, 95 (1955).
22. Longmire, C. and M. Rosenbluth, *Phys. Rev.*, **103**, 507 (1956).
23. Maecker, H., Th. Peters, and H. Schenck, *Z. Physik*, **140**, 119 (1955).
24. Marshak, R., *Ann. N. Y. Acad. Science*, **41**, 49 (1941).
25. Massey, H. S. W. and E. H. S. Burhop, *Electronic and Ionic Impact Phenomena*, Clarendon Press, Oxford, 1952, p. 88.
26. Oster, L., *Z. Astrophys.*, **47**, 169 (1959).
27. Pines, D. and D. Bohm, *Phys. Rev.*, **85**, 338 (1952).
28. Rand, S., *Phys. Fluids*, **4**, 1251 (1961).

29. Rosenbluth, M. N., W. MacDonald, and D. Judd, *Phys. Rev.*, **107**, 1 (1957).
30. Rosenbluth, M. N., and A. N. Kaufman, *Phys. Rev.*, **109**, 1 (1958).
31. Scheuer, P. A. G., *Monthly Notices, Roy. Astron. Soc. (London)*, **120**, 231, 1960.
32. Schwinger, J., *Phys. Rev.*, **75**, 1912 (1949).
33. Simon, A., *Phys. Rev.*, **100**, 1557 (1955).
34. Spitzer, L., *Monthly Notices, Roy. Astron. Soc. (London)*, **100**, 396 (1940).
35. Spitzer, L., *Astrophys. J.*, **116**, 299 (1952).
36. Spitzer, L., and R. Härm, *Phys. Rev.*, **89**, 977 (1953).
37. Trubnikov, B., *Doklady Akad. Nauk S.S.S.R.*, **118**, 913 (1958); *Soviet Phys. Doklady*, **3**, 136 (1958).
38. Westfold, K. C., *Astrophys. J.*, **130**, 241 (1959).

APPENDIX

The Boltzmann Equation

For precise results in the kinetic theory of gases, the Boltzmann equation must generally be employed. This relationship involves the quantity f, the density of particles in phase space, as a function of position **r** and velocity **w**.

More precisely, $f(\mathbf{r}, \mathbf{w}, t)dxdydzdw_x dw_y dw_z$ is the number of particles which lie within the spatial volume $dxdydz$, centered at **r**, and whose velocities lie within the intervals dw_x, dw_y, and dw_z centered at **w**. We define Df/Dt as the rate of change of f along the free trajectory of a particle, encounters between particles being ignored in the computation of this trajectory. The Boltzmann equation states that Df/Dt is entirely the result of encounters among the particles. For a group of identical particles this partial differential equation, discussed in detail by Chapman and Cowling (3), may be written in the form

$$\frac{\partial f}{\partial t} + \sum_i w_i \frac{\partial f}{\partial x_i} + \sum_i \frac{F_i}{m} \frac{\partial f}{\partial w_i} = \left(\frac{\partial f}{\partial t}\right)_{\text{coll}} \qquad (6\text{-}1)$$

where i takes the values 1, 2, and 3, with x_1, x_2, and x_3 representing the x, y, and z axes; w_i and F_i represent the ith components of the particle velocity and the external force, respectively, while m is the particle mass. The term $(\partial f/\partial t)_{\text{coll}}$ represents the change in f due to collisions between particles at a fixed point in space and time. Equation (6-1) is generally valid for conservative systems; **F** may include a magnetic component $q\mathbf{w} \times \mathbf{B}$. In the absence of collisions this equation reduces to Liouville's theorem, which states that for a conservative system f is constant along a dynamical trajectory.

The solution of equation (6-1) is rather complicated even in relatively simple cases. Here we shall show how this equation may be used to derive the basic macroscopic equations presented in Chapter 2. The macroscopic particle density, $n(\mathbf{r}, t)$, and the velocity, $\mathbf{v}(\mathbf{r}, t)$, may evidently be expressed in terms of f; we have

$$n(\mathbf{r}, t) = \iiint_{-\infty}^{+\infty} f(\mathbf{r}, \mathbf{w}, t)\,dw_x dw_y dw_z \tag{6-2}$$

$$\mathbf{v}(\mathbf{r}, t) = \frac{1}{n(\mathbf{r}, t)} \iiint_{-\infty}^{+\infty} \mathbf{w} f(\mathbf{r}, \mathbf{w}, t)\,dw_x dw_y dw_z \tag{6-3}$$

In general, the mean value $\overline{Q}(\mathbf{r}, t)$ for any quantity $Q(\mathbf{w})$ is given by

$$\overline{Q}(\mathbf{r}, t) = \frac{1}{n(\mathbf{r}, t)} \iiint_{-\infty}^{+\infty} Q(\mathbf{w}) f(\mathbf{r}, \mathbf{w}, t)\,dw_x dw_y dw_z \tag{6-4}$$

To obtain relations between macroscopic quantities we may integrate equation (6-1) over velocity space. We multiply this equation by $Q(\mathbf{w})dw_x dw_y dw_z$, where Q is some arbitrary function of $\mathbf{w}$, and integrate over all $\mathbf{w}$. In general we have

$$\begin{aligned} \iiint_{-\infty}^{+\infty} Q(\mathbf{w}) \frac{\partial f}{\partial t}\,dw_x dw_y dw_z &= \frac{\partial}{\partial t} \iiint_{-\infty}^{+\infty} Q(\mathbf{w}) f\,dw_x dw_y dw_z \\ &= \frac{\partial}{\partial t}(n\overline{Q}) \end{aligned} \tag{6-5}$$

Also

$$\begin{aligned} \iiint_{-\infty}^{+\infty} Q(\mathbf{w}) w_i \frac{\partial f}{\partial x_i}\,dw_x dw_y dw_z &= \frac{\partial}{\partial x_i} \iiint_{-\infty}^{+\infty} Q(\mathbf{w}) w_i f\,dw_x dw_y dw_z \\ &= \frac{\partial}{\partial x_i}(n\overline{w_i Q}) \end{aligned} \tag{6-6}$$

With an integration by parts over dw_i, we obtain

$$\iiint_{-\infty}^{+\infty} Q(\mathbf{w})F_i(\mathbf{r}, \mathbf{w}) \frac{\partial f}{\partial w_i} dw_x dw_y dw_z$$

$$= - \iiint_{-\infty}^{+\infty} f \frac{\partial}{\partial w_i} \{F_i(\mathbf{r}, \mathbf{w})Q(\mathbf{w})\} dw_x dw_y dw_z \quad (6\text{-}7)$$

$$= -n \overline{\frac{\partial}{\partial w_i} (F_i Q)}$$

since $f(\mathbf{w})$ is equal to zero for w_i equal to $\pm\infty$.

To obtain the equation of continuity we let Q equal 1. We may assume that $\partial F_i/\partial w_i$ equals zero; this relationship holds for magnetic forces, the only forces we shall consider which depend on the velocity. Moreover the integral of $(\partial f/\partial t)_{\text{coll}}$ over velocity space obviously vanishes, since collisions cannot change the total number of particles per cubic centimeter. Equations (6-5) and (6-6) then yield

$$\frac{\partial n}{\partial t} + \nabla \cdot (n\mathbf{v}) = 0 \quad (6\text{-}8)$$

The equation of momentum transfer, which in Chapter 2 has been called the equation of motion, is obtained by letting Q equal $m\mathbf{w}$. We obtain

$$\frac{\partial}{\partial t}(nm\mathbf{v}) + \nabla \cdot (nm\overline{\mathbf{w}\mathbf{w}}) - n\overline{\mathbf{F}} = \iiint_{-\infty}^{+\infty} m\mathbf{w} \left(\frac{\partial f}{\partial t}\right)_{\text{coll}} dw_x dw_y dw_z \quad (6\text{-}9)$$

This equation may be modified in several ways. The first term may be written in the form

$$\frac{\partial}{\partial t}(nm\mathbf{v}) = nm\frac{\partial \mathbf{v}}{\partial t} + \mathbf{v}\frac{\partial(nm)}{\partial t} \quad (6\text{-}10)$$

The quantity $\mathbf{ww}$ in the second term may be simplified if we let

$$\mathbf{w} = \mathbf{v} + \mathbf{u} \tag{6-11}$$

where $\mathbf{v}$ is the mean velocity, $\overline{\mathbf{w}}$, and $\mathbf{u}$ is the random velocity. Then we find

$$\nabla \cdot (nm\overline{\mathbf{ww}}) = \nabla \cdot (nm\mathbf{vv}) + \nabla \cdot (nm\overline{\mathbf{uu}}) \tag{6-12}$$

since $\overline{\mathbf{u}}$ vanishes. Comparison of equation (2-6) with equation (6-4) indicates that

$$nm\overline{\mathbf{uu}} = \mathbf{\Psi} \tag{6-13}$$

where $\mathbf{\Psi}$ is again the stress tensor. Expanding $\nabla \cdot (nm\mathbf{vv})$, we find that the second term in equation (6-9) becomes

$$\nabla \cdot (nm\overline{\mathbf{ww}}) = nm\mathbf{v} \cdot \nabla \mathbf{v} + \mathbf{v} \nabla \cdot (nm\mathbf{v}) + \nabla \cdot \mathbf{\Psi} \tag{6-14}$$

In the third term we may let

$$\mathbf{F} = q\mathbf{E} + q\mathbf{w} \times \mathbf{B} - m\nabla\phi \tag{6-15}$$

where $\mathbf{E}$ and $\mathbf{B}$ are the electric and magnetic field strengths in electromagnetic units, q is the particle charge in electromagnetic units, and ϕ is the gravitational potential. To obtain the mean value of $\mathbf{F}$, averaged over all the particles in a unit volume, we replace $\mathbf{w}$ by $\mathbf{v}$ in equation (6-15).

Finally we have the fourth term. This is obviously the momentum gained as a result of collisions by the particles in question. Collisions of identical particles with each other clearly produce no momentum gain. Collisions with other particles may yield a net momentum gain, which we denote by $\mathbf{P}$.

If these results for the four terms in equation (6-9) are now combined, and the equation of continuity (6-8) is also used, we now find

$$nm\left(\frac{\partial \mathbf{v}}{\partial t} + \mathbf{v} \cdot \nabla \mathbf{v}\right) = nq(\mathbf{E} + \mathbf{v} \times \mathbf{B}) - \nabla \cdot \mathbf{\Psi} - nm\nabla\phi + \mathbf{P} \tag{6-16}$$

This is the basic equation (2-4), which has been employed throughout this tract.

If we let Q equal $m\mathbf{ww}$ we obtain an equation for $\partial\mathbf{\Psi}/\partial t$, the time derivative of the stress tensor. This equation, which has been discussed by Jeans (4) and by a number of more recent authors (1), involves a tensor of third order, which is called the heat flow tensor. In simple cases as, for example, in a strong magnetic field, with conditions uniform along each line of force, this tensor is zero, and the change of $\mathbf{\Psi}$ with time may be computed. A systematic analysis of plasma dynamics, based on the equation for $\partial\mathbf{\Psi}/\partial t$ with the heat flow tensor set equal to zero, has been given by Buneman (2). In more general situations the flow of heat cannot be ignored, since it may have an important effect on the local velocity distribution and thus on the stress tensor. When the effect of heat flow cannot readily be evaluated, the macroscopic equations are not very useful, and the velocity distribution function $f(\mathbf{r}, \mathbf{w}, t)$ must be analyzed.

References

1. Bernstein, I. B., and S. K. Trehan, *Nuclear Fusion*, 1, 3 (1960), Section II.
2. Buneman, O., *Phys. Fluids*, 4, 669 (1961).
3. Chapman, S. and T. G. Cowling, *The Mathematical Theory of Non-Uniform Gases*, Cambridge University Press, Cambridge, 1953.
4. Jeans, J. H., *The Dynamical Theory of Gases*, 3rd ed., Cambridge University Press, Cambridge, 1921, Chapter IX.

Symbols

a — Radius of gyration, equation (1-4).

A — Atomic weight; A_e, A_i, atomic weight of electrons, positive ions.

A_D, A_r — Diffusion constant, equation (5-18), and recapture constant, equation (5-63).

$\mathbf{B}$ — Magnetic field strength, in gauss; $\delta\mathbf{B}$, change in $\mathbf{B}$ resulting from a displacement ξ, equation (4-8).

c — Velocity of light, 2.9979×10^{-10} cm/sec.

C — Numerical constant.

d — Attenuation distance for wave amplitude, equation (3-5).

e — Charge of proton, 4.803×10^{-10} e.s.u.
Base of Napierian logarithms.

$\mathbf{E}$ — Electric field strength, in e.m.u., equal to 10^8 times field strength in volts/cm; $E_{\|}$, $E_{\perp}$, components of $\mathbf{E}$ parallel and perpendicular to $\mathbf{B}$.

$\mathbf{E}_0$ — Maximum $\mathbf{E}$ in a sinusoidal wave.

$f(\mathbf{w})$ — Velocity distribution function; density of particles in velocity space; $f^{(0)}(w)$, Maxwell-Boltzmann distribution function, equation (5-8).

$\mathbf{F}$ — Force on a particle, in dynes.

g — Acceleration of gravity; $g_{\|}$, $g_{\perp}$, components of $\mathbf{g}$ parallel and perpendicular to $\mathbf{B}$.
Quantum mechanical correction factor; g_{ff}, correction factor for radiation from free-free transitions (bremsstrahlung), equations (5-59) and (5-61); g_{fn}, for radiation from free-bound transitions, equation (5-64).

$G(x)$ — Function defined in equation (5-20).

h — Debye shielding distance, equation (2-3).
Planck's constant, 6.625×10^{-27} gm cm^2/sec.

i — $(-1)^{1/2}$.

I — Total current, in e.m.u., equal to 1/10 times the current in amperes.

$I(x)$ — Imaginary part of x.

$\mathbf{j}$ — Current density, in e.m.u., equal to 1/10 times the current density in amp/cm^2.

k — Boltzmann constant, 1.380×10^{-16} erg/degree.

K Dielectric constant, equation (2-32).

$\mathcal{K}$ Coefficient of thermal conductivity; $\mathcal{K}_L$ thermal conductivity of a Lorentz gas, equation (5-47); $\mathcal{K}_\perp$, thermal conductivity transverse to a strong magnetic field, equation (5-53).

l Parameter characterizing Maxwell-Boltzmann distribution, equation (5-9).
Designation for left-handed circularly polarized wave.

$\ln x$ Natural, or Napierian, logarithm of x.

L Distance, length.

m Particle mass in grams; m_e and m_i, electron and ion mass.
Integer characterizing a particular normal mode, in which some perturbation varies as $\exp(im\theta)$.
Number of degrees of freedom.

M Reduced mass, equation (5-3).

n Particle density, per cm^3.

N No. of particles in a volume, or per linear cm (pinch effect).
Number of collisions experienced by a test particle.

o Designation for ordinary wave, **E** parallel to **B**.

p Pressure; p_e, p_i, pressure of electrons and positive ions; $p_\perp$, $p_\parallel$ pressure perpendicular and parallel to **B**.
Impact parameter, distance of closest approach in absence of interaction force.

p_0 Value of impact parameter such that deflection in orbital plane is $\pi/2$.

P_r Probability per unit time of electron runaway, equation (5-40).

$\mathbf{P}$ Rate of transfer of momentum, per cm^3 per sec, as a result of encounters with other particles; $\mathbf{P}_{ij}$, rate of momentum transfer to particles of type i from those of type j.

q Electrical charge, in e.m.u., equal to 1/10 times the charge in coulombs.

$\mathbf{Q}$ Heat flux; $\mathbf{Q}_\perp$, heat flux perpendicular to **B**.

$\mathbf{r}$ Position vector.

r Designation for right-handed circularly polarized wave.

R Reflection coefficient.
Radius of curvature of lines of force, in equations (1-12) and (1-13).

s Distance; ds, line element.

S Area; dS element of area.

t Time.

t_c Self-collision time, equation (5-26).

t_D Deflection time, equation (5-22).

t_E Energy exchange time, equation (5-25).

t_s Slowing-down time, equation (5-28).

t_{eq} Time of equipartition, between two groups of particles, equation (5-31).

T Temperature in degrees Kelvin, equal to 11,600 times kT in electron volts; T_e, T_i, kinetic temperature of electrons, positive ions; $T_\parallel$, $T_\perp$, kinetic temperature for velocities parallel and perpendicular to **B**.

u Relative velocity (Chapter 5).

u_m Root mean square maximum velocity (relative to wave) for particle trapping, equation (3-51).

U Electric potential, in e.m.u., equal to 10^8 times the potential in volts.

$\mathbf{v}$ Macroscopic velocity, equations (2-5) and (6-3); $\mathbf{v}_e$, $\mathbf{v}_i$, macroscopic velocity of electrons, positive ions.

$\mathbf{v}_D$ Diffusion velocity; $\mathbf{v}_{D\eta}$, $\mathbf{v}_{D\mu}$, $\mathbf{v}_{Dt}$, diffusion velocities resulting from finite resistivity (electron-ion collisions), viscosity (ion-ion collisions), and plasma turbulence, equations (2-39), (2-42), and (2-45), respectively.

V Volume; ΔV, dV, element of volume.

Phase velocity (Chapter 3).

V_A Alfvén velocity, equation (3-29).

V_S Sound velocity (acoustic or positive-ion waves), equation (3-21).

$\mathbf{w}$ Particle velocity; $w_\parallel$, $w_\perp$, velocity parallel and perpendicular to **B**, (Chapter 1); not to be confused with $\Delta w_\parallel$ and $\Delta w_\perp$ (Chapter 5).

$\mathbf{w}_D$ Drift velocity across magnetic field; velocity of the guiding center.

w_r Velocity of particles which experience resonant acceleration, equation (3-58).

W Energy; $\delta W(\xi, \xi)$, change in total energy, excluding kinetic, resulting from a displacement ξ; δW_S, δW_p, δW_v, contributions to δW resulting from changes at the plasma surface (equation 4-6), in the plasma volume (equation 4-7), and in the vacuum.

x, y, z Coordinate axes.

x Designation for extraordinary wave, **E** perpendicular to **B**.

Z Particle charge, in units of the proton charge; in all macroscopic equations, the average charge of the positive ions.

α Recombination coefficient, equation (5-67).

Fine structure constant, Section 5.2.

α, β Coefficients of thermoelectric effect, equations (5-43) and (5-44).

β Ratio of w to light velocity, c; $\beta_\parallel$ and $\beta_\perp$, corresponding ratios for $w_\parallel$, $w_\perp$.

$157{,}000^\circ\ Z^2/T$, equations (5-67) through (5-69).

γ	Ratio of sqecific heats.
	Euler's constant, epual to 0.5772.
γ_E, γ_{EB}	Factors by which η_L must be divided to give the actual η.
Γ	Parameter entering the probability of electron runaway, equation (5-39), about equal to the ratio of mean electron drift velocity to random thermal velocity.
δ_T	Factor by which $\mathcal{K}_L$ must be multiplied to give $\mathcal{K}$.
Δ	Increment; $\Delta w_{\parallel}$, $\Delta w_{\perp}$, increment of **w** parallel and perpendicular to **w**.
ϵ	Factor by which $\mathcal{K}$ must be reduced because of thermoelectric effect.
ϵ_{ff}	Rate of radiation in free-free transitions, equation (5-60).
ϵ_s	Rate of synchrotron radiation, equation (5-73); ϵ_{sn}, radiation rate for *n*th harmonic, equation (5-70).
ζ	Ratio of values of $\omega_p{}^2$ for two components in a plasma.
η	Resistivity, in e.m.u., equal to 10^9 times resistivity in ohm-cm; η_L, resistivity of a Lorentz gas, equation (5-34); $\eta_\perp$, resistivity transverse to a strong magnetic field, equation (5-42).
θ	Angle.
κ	Wave number, $2\pi/\lambda$.
κ_ν	Absorption coefficient per cm^3 for radiation of frequency ν; units are cm^2/cm^3.
λ	Wavelength.
	Thermoelectric coefficient in a strong magnetic field, equation (5-52).
Λ	Ratio of Debye shielding distance to p_0, equation (5-14).
μ	Magnetic moment (diamagnetic) of a charged particle gyrating about magnetic lines of force, equation (1-17).
	Coefficient of viscosity; $\mu_\perp$, coefficient of viscosity for shearing stresses in plane perpendicular to **B**.
ν	Frequency, including collision frequency; ν_c cyclotron frequency.
$\xi(\mathbf{r})$	Arbitrary displacement in a fluid.
ρ	Mass density, in $grams/cm^3$.
σ	Charge density, in e.m.u., per cm^3, equal to 1/10 times charge density in $coulombs/cm^3$.
	Decay rate of wave amplitude.
σ_s	Cross section of free electron for scattering a photon, equation (5-56).
σ_{cn}	Cross section of a bare nucleus for capturing a free electron into level *n*, equation (5-62).
τ	Decay time for magnetic field.

ϕ	Gravitational potential.
$\phi(\beta)$	Function defined in equation (5-69).
Φ	Flux through surface.
$\Phi(x)$	Error function.
χ	Deflection angle in orbital plane, in encounter between two particles.
ψ	Angle characterizing encounter between two particles, Figure 5.1 and equation (5-1).
$\mathbf{\Psi}$	Stress tensor, equations (2-6) and (6-13).
ω	Angular frequency; ω_c, cyclotron frequency; ω_{ce}, ω_{ci}, cyclotron frequency of electrons, positive ions.
ω_p	Plasma frequency, equation (3-8).
$\omega_r(w_r)$	Parameter measuring the number of particles with the velocity w_r, equation (3-60).
ω_t	Oscillation frequency of trapped particles, equation (3-51).
Ω	Solid angle; $d\Omega$, element of solid angle.
$\mathbf{\nabla}$	Gradient; $\nabla_\parallel$, $\nabla_\perp$, gradient parallel and perpendicular to **B**.
$\langle X \rangle$	Diffusion coefficient; the value of X summed over all encounters experienced by a test particle per second averaged over all test particles within a volume element in phase space.

Index

A CATALOG OF SELECTED

DOVER BOOKS

IN SCIENCE AND MATHEMATICS

Astronomy

CHARIOTS FOR APOLLO: The NASA History of Manned Lunar Spacecraft to 1969, Courtney G. Brooks, James M. Grimwood, and Loyd S. Swenson, Jr. This illustrated history by a trio of experts is the definitive reference on the Apollo spacecraft and lunar modules. It traces the vehicles' design, development, and operation in space. More than 100 photographs and illustrations. 576pp. 6 3/4 x 9 1/4. 0-486-46756-2

EXPLORING THE MOON THROUGH BINOCULARS AND SMALL TELESCOPES, Ernest H. Cherrington, Jr. Informative, profusely illustrated guide to locating and identifying craters, rills, seas, mountains, other lunar features. Newly revised and updated with special section of new photos. Over 100 photos and diagrams. 240pp. 8 1/4 x 11. 0-486-24491-1

WHERE NO MAN HAS GONE BEFORE: A History of NASA's Apollo Lunar Expeditions, William David Compton. Introduction by Paul Dickson. This official NASA history traces behind-the-scenes conflicts and cooperation between scientists and engineers. The first half concerns preparations for the Moon landings, and the second half documents the flights that followed Apollo 11. 1989 edition. 432pp. 7 x 10. 0-486-47888-2

APOLLO EXPEDITIONS TO THE MOON: The NASA History, Edited by Edgar M. Cortright. Official NASA publication marks the 40th anniversary of the first lunar landing and features essays by project participants recalling engineering and administrative challenges. Accessible, jargon-free accounts, highlighted by numerous illustrations. 336pp. 8 3/8 x 10 7/8. 0-486-47175-6

ON MARS: Exploration of the Red Planet, 1958-1978--The NASA History, Edward Clinton Ezell and Linda Neuman Ezell. NASA's official history chronicles the start of our explorations of our planetary neighbor. It recounts cooperation among government, industry, and academia, and it features dozens of photos from Viking cameras. 560pp. 6 3/4 x 9 1/4. 0-486-46757-0

ARISTARCHUS OF SAMOS: The Ancient Copernicus, Sir Thomas Heath. Heath's history of astronomy ranges from Homer and Hesiod to Aristarchus and includes quotes from numerous thinkers, compilers, and scholasticists from Thales and Anaximander through Pythagoras, Plato, Aristotle, and Heraclides. 34 figures. 448pp. 5 3/8 x 8 1/2. 0-486-43886-4

AN INTRODUCTION TO CELESTIAL MECHANICS, Forest Ray Moulton. Classic text still unsurpassed in presentation of fundamental principles. Covers rectilinear motion, central forces, problems of two and three bodies, much more. Includes over 200 problems, some with answers. 437pp. 5 3/8 x 8 1/2. 0-486-64687-4

BEYOND THE ATMOSPHERE: Early Years of Space Science, Homer E. Newell. This exciting survey is the work of a top NASA administrator who chronicles technological advances, the relationship of space science to general science, and the space program's social, political, and economic contexts. 528pp. 6 3/4 x 9 1/4. 0-486-47464-X

STAR LORE: Myths, Legends, and Facts, William Tyler Olcott. Captivating retellings of the origins and histories of ancient star groups include Pegasus, Ursa Major, Pleiades, signs of the zodiac, and other constellations. "Classic." – *Sky & Telescope.* 58 illustrations. 544pp. 5 3/8 x 8 1/2. 0-486-43581-4

A COMPLETE MANUAL OF AMATEUR ASTRONOMY: Tools and Techniques for Astronomical Observations, P. Clay Sherrod with Thomas L. Koed. Concise, highly readable book discusses the selection, set-up, and maintenance of a telescope; amateur studies of the sun; lunar topography and occultations; and more. 124 figures. 26 halftones. 37 tables. 335pp. 6 1/2 x 9 1/4. 0-486-42820-6

Chemistry

MOLECULAR COLLISION THEORY, M. S. Child. This high-level monograph offers an analytical treatment of classical scattering by a central force, quantum scattering by a central force, elastic scattering phase shifts, and semi-classical elastic scattering. 1974 edition. 310pp. 5 3/8 x 8 1/2. 0-486-69437-2

HANDBOOK OF COMPUTATIONAL QUANTUM CHEMISTRY, David B. Cook. This comprehensive text provides upper-level undergraduates and graduate students with an accessible introduction to the implementation of quantum ideas in molecular modeling, exploring practical applications alongside theoretical explanations. 1998 edition. 832pp. 5 3/8 x 8 1/2. 0-486-44307-8

RADIOACTIVE SUBSTANCES, Marie Curie. The celebrated scientist's thesis, which directly preceded her 1903 Nobel Prize, discusses establishing atomic character of radioactivity; extraction from pitchblende of polonium and radium; isolation of pure radium chloride; more. 96pp. 5 3/8 x 8 1/2. 0-486-42550-9

CHEMICAL MAGIC, Leonard A. Ford. Classic guide provides intriguing entertainment while elucidating sound scientific principles, with more than 100 unusual stunts: cold fire, dust explosions, a nylon rope trick, a disappearing beaker, much more. 128pp. 5 3/8 x 8 1/2. 0-486-67628-5

ALCHEMY, E. J. Holmyard. Classic study by noted authority covers 2,000 years of alchemical history: religious, mystical overtones; apparatus; signs, symbols, and secret terms; advent of scientific method, much more. Illustrated. 320pp. 5 3/8 x 8 1/2. 0-486-26298-7

CHEMICAL KINETICS AND REACTION DYNAMICS, Paul L. Houston. This text teaches the principles underlying modern chemical kinetics in a clear, direct fashion, using several examples to enhance basic understanding. Solutions to selected problems. 2001 edition. 352pp. 8 3/8 x 11. 0-486-45334-0

PROBLEMS AND SOLUTIONS IN QUANTUM CHEMISTRY AND PHYSICS, Charles S. Johnson and Lee G. Pedersen. Unusually varied problems, with detailed solutions, cover of quantum mechanics, wave mechanics, angular momentum, molecular spectroscopy, scattering theory, more. 280 problems, plus 139 supplementary exercises. 430pp. 6 1/2 x 9 1/4. 0-486-65236-X

ELEMENTS OF CHEMISTRY, Antoine Lavoisier. Monumental classic by the founder of modern chemistry features first explicit statement of law of conservation of matter in chemical change, and more. Facsimile reprint of original (1790) Kerr translation. 539pp. 5 3/8 x 8 1/2. 0-486-64624-6

MAGNETISM AND TRANSITION METAL COMPLEXES, F. E. Mabbs and D. J. Machin. A detailed view of the calculation methods involved in the magnetic properties of transition metal complexes, this volume offers sufficient background for original work in the field. 1973 edition. 240pp. 5 3/8 x 8 1/2. 0-486-46284-6

GENERAL CHEMISTRY, Linus Pauling. Revised third edition of classic first-year text by Nobel laureate. Atomic and molecular structure, quantum mechanics, statistical mechanics, thermodynamics correlated with descriptive chemistry. Problems. 992pp. 5 3/8 x 8 1/2. 0-486-65622-5

ELECTROLYTE SOLUTIONS: Second Revised Edition, R. A. Robinson and R. H. Stokes. Classic text deals primarily with measurement, interpretation of conductance, chemical potential, and diffusion in electrolyte solutions. Detailed theoretical interpretations, plus extensive tables of thermodynamic and transport properties. 1970 edition. 590pp. 5 3/8 x 8 1/2. 0-486-42225-9

Engineering

FUNDAMENTALS OF ASTRODYNAMICS, Roger R. Bate, Donald D. Mueller, and Jerry E. White. Teaching text developed by U.S. Air Force Academy develops the basic two-body and n-body equations of motion; orbit determination; classical orbital elements, coordinate transformations; differential correction; more. 1971 edition. 455pp. 5 3/8 x 8 1/2. 0-486-60061-0

INTRODUCTION TO CONTINUUM MECHANICS FOR ENGINEERS: Revised Edition, Ray M. Bowen. This self-contained text introduces classical continuum models within a modern framework. Its numerous exercises illustrate the governing principles, linearizations, and other approximations that constitute classical continuum models. 2007 edition. 320pp. 6 1/8 x 9 1/4. 0-486-47460-7

ENGINEERING MECHANICS FOR STRUCTURES, Louis L. Bucciarelli. This text explores the mechanics of solids and statics as well as the strength of materials and elasticity theory. Its many design exercises encourage creative initiative and systems thinking. 2009 edition. 320pp. 6 1/8 x 9 1/4. 0-486-46855-0

FEEDBACK CONTROL THEORY, John C. Doyle, Bruce A. Francis and Allen R. Tannenbaum. This excellent introduction to feedback control system design offers a theoretical approach that captures the essential issues and can be applied to a wide range of practical problems. 1992 edition. 224pp. 6 1/2 x 9 1/4. 0-486-46933-6

THE FORCES OF MATTER, Michael Faraday. These lectures by a famous inventor offer an easy-to-understand introduction to the interactions of the universe's physical forces. Six essays explore gravitation, cohesion, chemical affinity, heat, magnetism, and electricity. 1993 edition. 96pp. 5 3/8 x 8 1/2. 0-486-47482-8

DYNAMICS, Lawrence E. Goodman and William H. Warner. Beginning engineering text introduces calculus of vectors, particle motion, dynamics of particle systems and plane rigid bodies, technical applications in plane motions, and more. Exercises and answers in every chapter. 619pp. 5 3/8 x 8 1/2. 0-486-42006-X

ADAPTIVE FILTERING PREDICTION AND CONTROL, Graham C. Goodwin and Kwai Sang Sin. This unified survey focuses on linear discrete-time systems and explores natural extensions to nonlinear systems. It emphasizes discrete-time systems, summarizing theoretical and practical aspects of a large class of adaptive algorithms. 1984 edition. 560pp. 6 1/2 x 9 1/4. 0-486-46932-8

INDUCTANCE CALCULATIONS, Frederick W. Grover. This authoritative reference enables the design of virtually every type of inductor. It features a single simple formula for each type of inductor, together with tables containing essential numerical factors. 1946 edition. 304pp. 5 3/8 x 8 1/2. 0-486-47440-2

THERMODYNAMICS: Foundations and Applications, Elias P. Gyftopoulos and Gian Paolo Beretta. Designed by two MIT professors, this authoritative text discusses basic concepts and applications in detail, emphasizing generality, definitions, and logical consistency. More than 300 solved problems cover realistic energy systems and processes. 800pp. 6 1/8 x 9 1/4. 0-486-43932-1

THE FINITE ELEMENT METHOD: Linear Static and Dynamic Finite Element Analysis, Thomas J. R. Hughes. Text for students without in-depth mathematical training, this text includes a comprehensive presentation and analysis of algorithms of time-dependent phenomena plus beam, plate, and shell theories. Solution guide available upon request. 672pp. 6 1/2 x 9 1/4. 0-486-41181-8

HELICOPTER THEORY, Wayne Johnson. Monumental engineering text covers vertical flight, forward flight, performance, mathematics of rotating systems, rotary wing dynamics and aerodynamics, aeroelasticity, stability and control, stall, noise, and more. 189 illustrations. 1980 edition. 1089pp. 5 5/8 x 8 1/4. 0-486-68230-7

MATHEMATICAL HANDBOOK FOR SCIENTISTS AND ENGINEERS: Definitions, Theorems, and Formulas for Reference and Review, Granino A. Korn and Theresa M. Korn. Convenient access to information from every area of mathematics: Fourier transforms, Z transforms, linear and nonlinear programming, calculus of variations, random-process theory, special functions, combinatorial analysis, game theory, much more. 1152pp. 5 3/8 x 8 1/2. 0-486-41147-8

A HEAT TRANSFER TEXTBOOK: Fourth Edition, John H. Lienhard V and John H. Lienhard IV. This introduction to heat and mass transfer for engineering students features worked examples and end-of-chapter exercises. Worked examples and end-of-chapter exercises appear throughout the book, along with well-drawn, illuminating figures. 768pp. 7 x 9 1/4. 0-486-47931-5

BASIC ELECTRICITY, U.S. Bureau of Naval Personnel. Originally a training course; best nontechnical coverage. Topics include batteries, circuits, conductors, AC and DC, inductance and capacitance, generators, motors, transformers, amplifiers, etc. Many questions with answers. 349 illustrations. 1969 edition. 448pp. 6 1/2 x 9 1/4. 0-486-20973-3

BASIC ELECTRONICS, U.S. Bureau of Naval Personnel. Clear, well-illustrated introduction to electronic equipment covers numerous essential topics: electron tubes, semiconductors, electronic power supplies, tuned circuits, amplifiers, receivers, ranging and navigation systems, computers, antennas, more. 560 illustrations. 567pp. 6 1/2 x 9 1/4. 0-486-21076-6

BASIC WING AND AIRFOIL THEORY, Alan Pope. This self-contained treatment by a pioneer in the study of wind effects covers flow functions, airfoil construction and pressure distribution, finite and monoplane wings, and many other subjects. 1951 edition. 320pp. 5 3/8 x 8 1/2. 0-486-47188-8

SYNTHETIC FUELS, Ronald F. Probstein and R. Edwin Hicks. This unified presentation examines the methods and processes for converting coal, oil, shale, tar sands, and various forms of biomass into liquid, gaseous, and clean solid fuels. 1982 edition. 512pp. 6 1/8 x 9 1/4. 0-486-44977-7

THEORY OF ELASTIC STABILITY, Stephen P. Timoshenko and James M. Gere. Written by world-renowned authorities on mechanics, this classic ranges from theoretical explanations of 2- and 3-D stress and strain to practical applications such as torsion, bending, and thermal stress. 1961 edition. 560pp. 5 3/8 x 8 1/2. 0-486-47207-8

PRINCIPLES OF DIGITAL COMMUNICATION AND CODING, Andrew J. Viterbi and Jim K. Omura. This classic by two digital communications experts is geared toward students of communications theory and to designers of channels, links, terminals, modems, or networks used to transmit and receive digital messages. 1979 edition. 576pp. 6 1/8 x 9 1/4. 0-486-46901-8

LINEAR SYSTEM THEORY: The State Space Approach, Lotfi A. Zadeh and Charles A. Desoer. Written by two pioneers in the field, this exploration of the state space approach focuses on problems of stability and control, plus connections between this approach and classical techniques. 1963 edition. 656pp. 6 1/8 x 9 1/4. 0-486-46663-9

Mathematics–Bestsellers

HANDBOOK OF MATHEMATICAL FUNCTIONS: with Formulas, Graphs, and Mathematical Tables, Edited by Milton Abramowitz and Irene A. Stegun. A classic resource for working with special functions, standard trig, and exponential logarithmic definitions and extensions, it features 29 sets of tables, some to as high as 20 places. 1046pp. 8 x 10 1/2. 0-486-61272-4

ABSTRACT AND CONCRETE CATEGORIES: The Joy of Cats, Jiri Adamek, Horst Herrlich, and George E. Strecker. This up-to-date introductory treatment employs category theory to explore the theory of structures. Its unique approach stresses concrete categories and presents a systematic view of factorization structures. Numerous examples. 1990 edition, updated 2004. 528pp. 6 1/8 x 9 1/4. 0-486-46934-4

MATHEMATICS: Its Content, Methods and Meaning, A. D. Aleksandrov, A. N. Kolmogorov, and M. A. Lavrent'ev. Major survey offers comprehensive, coherent discussions of analytic geometry, algebra, differential equations, calculus of variations, functions of a complex variable, prime numbers, linear and non-Euclidean geometry, topology, functional analysis, more. 1963 edition. 1120pp. 5 3/8 x 8 1/2. 0-486-40916-3

INTRODUCTION TO VECTORS AND TENSORS: Second Edition--Two Volumes Bound as One, Ray M. Bowen and C.-C. Wang. Convenient single-volume compilation of two texts offers both introduction and in-depth survey. Geared toward engineering and science students rather than mathematicians, it focuses on physics and engineering applications. 1976 edition. 560pp. 6 1/2 x 9 1/4. 0-486-46914-X

AN INTRODUCTION TO ORTHOGONAL POLYNOMIALS, Theodore S. Chihara. Concise introduction covers general elementary theory, including the representation theorem and distribution functions, continued fractions and chain sequences, the recurrence formula, special functions, and some specific systems. 1978 edition. 272pp. 5 3/8 x 8 1/2. 0-486-47929-3

ADVANCED MATHEMATICS FOR ENGINEERS AND SCIENTISTS, Paul DuChateau. This primary text and supplemental reference focuses on linear algebra, calculus, and ordinary differential equations. Additional topics include partial differential equations and approximation methods. Includes solved problems. 1992 edition. 400pp. 7 1/2 x 9 1/4. 0-486-47930-7

PARTIAL DIFFERENTIAL EQUATIONS FOR SCIENTISTS AND ENGINEERS, Stanley J. Farlow. Practical text shows how to formulate and solve partial differential equations. Coverage of diffusion-type problems, hyperbolic-type problems, elliptic-type problems, numerical and approximate methods. Solution guide available upon request. 1982 edition. 414pp. 6 1/8 x 9 1/4. 0-486-67620-X

VARIATIONAL PRINCIPLES AND FREE-BOUNDARY PROBLEMS, Avner Friedman. Advanced graduate-level text examines variational methods in partial differential equations and illustrates their applications to free-boundary problems. Features detailed statements of standard theory of elliptic and parabolic operators. 1982 edition. 720pp. 6 1/8 x 9 1/4. 0-486-47853-X

LINEAR ANALYSIS AND REPRESENTATION THEORY, Steven A. Gaal. Unified treatment covers topics from the theory of operators and operator algebras on Hilbert spaces; integration and representation theory for topological groups; and the theory of Lie algebras, Lie groups, and transform groups. 1973 edition. 704pp. 6 1/8 x 9 1/4. 0-486-47851-3

A SURVEY OF INDUSTRIAL MATHEMATICS, Charles R. MacCluer. Students learn how to solve problems they'll encounter in their professional lives with this concise single-volume treatment. It employs MATLAB and other strategies to explore typical industrial problems. 2000 edition. 384pp. 5 3/8 x 8 1/2. 0-486-47702-9

NUMBER SYSTEMS AND THE FOUNDATIONS OF ANALYSIS, Elliott Mendelson. Geared toward undergraduate and beginning graduate students, this study explores natural numbers, integers, rational numbers, real numbers, and complex numbers. Numerous exercises and appendixes supplement the text. 1973 edition. 368pp. 5 3/8 x 8 1/2. 0-486-45792-3

A FIRST LOOK AT NUMERICAL FUNCTIONAL ANALYSIS, W. W. Sawyer. Text by renowned educator shows how problems in numerical analysis lead to concepts of functional analysis. Topics include Banach and Hilbert spaces, contraction mappings, convergence, differentiation and integration, and Euclidean space. 1978 edition. 208pp. 5 3/8 x 8 1/2. 0-486-47882-3

FRACTALS, CHAOS, POWER LAWS: Minutes from an Infinite Paradise, Manfred Schroeder. A fascinating exploration of the connections between chaos theory, physics, biology, and mathematics, this book abounds in award-winning computer graphics, optical illusions, and games that clarify memorable insights into self-similarity. 1992 edition. 448pp. 6 1/8 x 9 1/4. 0-486-47204-3

SET THEORY AND THE CONTINUUM PROBLEM, Raymond M. Smullyan and Melvin Fitting. A lucid, elegant, and complete survey of set theory, this three-part treatment explores axiomatic set theory, the consistency of the continuum hypothesis, and forcing and independence results. 1996 edition. 336pp. 6 x 9. 0-486-47484-4

DYNAMICAL SYSTEMS, Shlomo Sternberg. A pioneer in the field of dynamical systems discusses one-dimensional dynamics, differential equations, random walks, iterated function systems, symbolic dynamics, and Markov chains. Supplementary materials include PowerPoint slides and MATLAB exercises. 2010 edition. 272pp. 6 1/8 x 9 1/4. 0-486-47705-3

ORDINARY DIFFERENTIAL EQUATIONS, Morris Tenenbaum and Harry Pollard. Skillfully organized introductory text examines origin of differential equations, then defines basic terms and outlines general solution of a differential equation. Explores integrating factors; dilution and accretion problems; Laplace Transforms; Newton's Interpolation Formulas, more. 818pp. 5 3/8 x 8 1/2. 0-486-64940-7

MATROID THEORY, D. J. A. Welsh. Text by a noted expert describes standard examples and investigation results, using elementary proofs to develop basic matroid properties before advancing to a more sophisticated treatment. Includes numerous exercises. 1976 edition. 448pp. 5 3/8 x 8 1/2. 0-486-47439-9

THE CONCEPT OF A RIEMANN SURFACE, Hermann Weyl. This classic on the general history of functions combines function theory and geometry, forming the basis of the modern approach to analysis, geometry, and topology. 1955 edition. 208pp. 5 3/8 x 8 1/2. 0-486-47004-0

THE LAPLACE TRANSFORM, David Vernon Widder. This volume focuses on the Laplace and Stieltjes transforms, offering a highly theoretical treatment. Topics include fundamental formulas, the moment problem, monotonic functions, and Tauberian theorems. 1941 edition. 416pp. 5 3/8 x 8 1/2. 0-486-47755-X

Mathematics–Logic and Problem Solving

PERPLEXING PUZZLES AND TANTALIZING TEASERS, Martin Gardner. Ninety-three riddles, mazes, illusions, tricky questions, word and picture puzzles, and other challenges offer hours of entertainment for youngsters. Filled with rib-tickling drawings. Solutions. 224pp. 5 3/8 x 8 1/2. 0-486-25637-5

MY BEST MATHEMATICAL AND LOGIC PUZZLES, Martin Gardner. The noted expert selects 70 of his favorite "short" puzzles. Includes The Returning Explorer, The Mutilated Chessboard, Scrambled Box Tops, and dozens more. Complete solutions included. 96pp. 5 3/8 x 8 1/2. 0-486-28152-3

THE LADY OR THE TIGER?: and Other Logic Puzzles, Raymond M. Smullyan. Created by a renowned puzzle master, these whimsically themed challenges involve paradoxes about probability, time, and change; metapuzzles; and self-referentiality. Nineteen chapters advance in difficulty from relatively simple to highly complex. 1982 edition. 240pp. 5 3/8 x 8 1/2. 0-486-47027-X

SATAN, CANTOR AND INFINITY: Mind-Boggling Puzzles, Raymond M. Smullyan. A renowned mathematician tells stories of knights and knaves in an entertaining look at the logical precepts behind infinity, probability, time, and change. Requires a strong background in mathematics. Complete solutions. 288pp. 5 3/8 x 8 1/2. 0-486-47036-9

THE RED BOOK OF MATHEMATICAL PROBLEMS, Kenneth S. Williams and Kenneth Hardy. Handy compilation of 100 practice problems, hints and solutions indispensable for students preparing for the William Lowell Putnam and other mathematical competitions. Preface to the First Edition. Sources. 1988 edition. 192pp. 5 3/8 x 8 1/2. 0-486-69415-1

KING ARTHUR IN SEARCH OF HIS DOG AND OTHER CURIOUS PUZZLES, Raymond M. Smullyan. This fanciful, original collection for readers of all ages features arithmetic puzzles, logic problems related to crime detection, and logic and arithmetic puzzles involving King Arthur and his Dogs of the Round Table. 160pp. 5 3/8 x 8 1/2. 0-486-47435-6

UNDECIDABLE THEORIES: Studies in Logic and the Foundation of Mathematics, Alfred Tarski in collaboration with Andrzej Mostowski and Raphael M. Robinson. This well-known book by the famed logician consists of three treatises: "A General Method in Proofs of Undecidability," "Undecidability and Essential Undecidability in Mathematics," and "Undecidability of the Elementary Theory of Groups." 1953 edition. 112pp. 5 3/8 x 8 1/2. 0-486-47703-7

LOGIC FOR MATHEMATICIANS, J. Barkley Rosser. Examination of essential topics and theorems assumes no background in logic. "Undoubtedly a major addition to the literature of mathematical logic." – *Bulletin of the American Mathematical Society.* 1978 edition. 592pp. 6 1/8 x 9 1/4. 0-486-46898-4

INTRODUCTION TO PROOF IN ABSTRACT MATHEMATICS, Andrew Wohlgemuth. This undergraduate text teaches students what constitutes an acceptable proof, and it develops their ability to do proofs of routine problems as well as those requiring creative insights. 1990 edition. 384pp. 6 1/2 x 9 1/4. 0-486-47854-8

FIRST COURSE IN MATHEMATICAL LOGIC, Patrick Suppes and Shirley Hill. Rigorous introduction is simple enough in presentation and context for wide range of students. Symbolizing sentences; logical inference; truth and validity; truth tables; terms, predicates, universal quantifiers; universal specification and laws of identity; more. 288pp. 5 3/8 x 8 1/2. 0-486-42259-3

Mathematics-Algebra and Calculus

VECTOR CALCULUS, Peter Baxandall and Hans Liebeck. This introductory text offers a rigorous, comprehensive treatment. Classical theorems of vector calculus are amply illustrated with figures, worked examples, physical applications, and exercises with hints and answers. 1986 edition. 560pp. 5 3/8 x 8 1/2. 0-486-46620-5

ADVANCED CALCULUS: An Introduction to Classical Analysis, Louis Brand. A course in analysis that focuses on the functions of a real variable, this text introduces the basic concepts in their simplest setting and illustrates its teachings with numerous examples, theorems, and proofs. 1955 edition. 592pp. 5 3/8 x 8 1/2. 0-486-44548-8

ADVANCED CALCULUS, Avner Friedman. Intended for students who have already completed a one-year course in elementary calculus, this two-part treatment advances from functions of one variable to those of several variables. Solutions. 1971 edition. 432pp. 5 3/8 x 8 1/2. 0-486-45795-8

METHODS OF MATHEMATICS APPLIED TO CALCULUS, PROBABILITY, AND STATISTICS, Richard W. Hamming. This 4-part treatment begins with algebra and analytic geometry and proceeds to an exploration of the calculus of algebraic functions and transcendental functions and applications. 1985 edition. Includes 310 figures and 18 tables. 880pp. 6 1/2 x 9 1/4. 0-486-43945-3

BASIC ALGEBRA I: Second Edition, Nathan Jacobson. A classic text and standard reference for a generation, this volume covers all undergraduate algebra topics, including groups, rings, modules, Galois theory, polynomials, linear algebra, and associative algebra. 1985 edition. 528pp. 6 1/8 x 9 1/4. 0-486-47189-6

BASIC ALGEBRA II: Second Edition, Nathan Jacobson. This classic text and standard reference comprises all subjects of a first-year graduate-level course, including in-depth coverage of groups and polynomials and extensive use of categories and functors. 1989 edition. 704pp. 6 1/8 x 9 1/4. 0-486-47187-X

CALCULUS: An Intuitive and Physical Approach (Second Edition), Morris Kline. Application-oriented introduction relates the subject as closely as possible to science with explorations of the derivative; differentiation and integration of the powers of x; theorems on differentiation, antidifferentiation; the chain rule; trigonometric functions; more. Examples. 1967 edition. 960pp. 6 1/2 x 9 1/4. 0-486-40453-6

ABSTRACT ALGEBRA AND SOLUTION BY RADICALS, John E. Maxfield and Margaret W. Maxfield. Accessible advanced undergraduate-level text starts with groups, rings, fields, and polynomials and advances to Galois theory, radicals and roots of unity, and solution by radicals. Numerous examples, illustrations, exercises, appendixes. 1971 edition. 224pp. 6 1/8 x 9 1/4. 0-486-47723-1

AN INTRODUCTION TO THE THEORY OF LINEAR SPACES, Georgi E. Shilov. Translated by Richard A. Silverman. Introductory treatment offers a clear exposition of algebra, geometry, and analysis as parts of an integrated whole rather than separate subjects. Numerous examples illustrate many different fields, and problems include hints or answers. 1961 edition. 320pp. 5 3/8 x 8 1/2. 0-486-63070-6

LINEAR ALGEBRA, Georgi E. Shilov. Covers determinants, linear spaces, systems of linear equations, linear functions of a vector argument, coordinate transformations, the canonical form of the matrix of a linear operator, bilinear and quadratic forms, and more. 387pp. 5 3/8 x 8 1/2. 0-486-63518-X